FORD FE Engines

HOW TO REBUILD

Barry Rabotnick

CarTech®

CarTech®

CarTech®, Inc.
6118 Main St.,
North Branch, MN 55056
Phone: 651-277-1200 or 800-551-4754
Fax: 651-277-1203
www.cartechbooks.com

Edit by Bob Wilson
Layout by Monica Seiberlich

ISBN 978-1-61325-244-4
Item No. SA352

Library of Congress Cataloging-in-Publication Data

Names: Rabotnick, Barry, author.
Title: Ford FE engines : how to rebuild / Barry Rabotnick.
Description: Forest Lake, MN : CarTech Books, [2018]
Identifiers: LCCN 2017051151 | ISBN 9781613252444
Subjects: LCSH: Ford automobile–Motors–Maintenance and
    repair–Handbooks, manuals, etc. | Automobiles
    –Motors–Maintenance and repair–Handbooks, manuals, etc. |
    LCGFT: Handbooks and manuals.
Classification: LCC TL215.F7 R32 2018 | DDC 629.25/040288–dc23
LC record available at https://lccn.loc.gov/2017051151

Written, edited, and designed in the U.S.A.
Printed in China
10 9 8 7

DISTRIBUTION BY:

Europe
PGUK
63 Hatton Garden
London EC1N 8LE, England
Phone: 020 7061 1980 • Fax: 020 7242 3725
www.pguk.co.uk

Australia
Renniks Publications Ltd.
3/37-39 Green Street
Banksmeadow, NSW 2109, Australia
Phone: 2 9695 7055 • Fax: 2 9695 7355
www.renniks.com

Canada
Login Canada
300 Saulteaux Crescent
Winnipeg, MB, R3J-3T2 Canada
Phone: 800 665 1148 • Fax: 800 665 0103
www.lb.ca

CONTENTS

This book took a great deal of work and significant time to complete. Many folks helped put it together, and I want to make certain that they receive credit for their efforts and contributions. First is to Bob Wilson from CarTech. Without his efforts and tenacity, this publication would never have been completed. He will probably edit this out, but I hope he doesn't because his efforts are very much appreciated and his recognition is highly deserved.

The folks I work with at Survival Motorsports put up with a lot during this project. Stopping a complex process in order to take photos and write out details is a significant disruption to an engine builder's daily routine. William Blair, our lead machinist, made a huge contribution in terms of posing machining and assembly operations throughout the build. Brian Blair and Nancy Schultz helped keep the proverbial "wheels on the bus" in the shop while I was focused on the book. Marc Weiner, a former employee and longtime friend, also helped with images and with teardown and assembly assistance on the engine we featured. I cannot forget to mention Mr. Pankow, as it is his 428 CJ and 1969 Shelby GT500 that modeled for most of the images we used.

Every author thanks his or her family and I am no exception to that rule. My wife, Susan, and my three daughters, Autumn, Summer, and Lily, have spent the past months and years hearing more about Ford FE engines than they ever expected or desired to. In between their dance lessons, equestrian events, and music activities, they have learned about crankshafts, cylinder heads, and intake manifolds.

The FE Ford engine is blessed with a strong and enthusiastic following on the Internet. The Ford FE forum, Jay Brown's FE Power forum, and a Wes Adams's FE Fanatics Facebook forum are all great places to invest time before starting out on an FE engine build. Numerous folks are willing to freely trade historical information and firsthand experience in every facet, from racing to original production data to proven repair techniques. Without their help and shared knowledge, this engine would not be as popular in the market as it is today. As it is, the FE enthusiast will benefit from one of the best cases of documented cumulative history and assembled knowledge of any automotive engine new or old.

The engine featured in this particular book is mostly original. But the FE engine has greatly benefited from the aftermarket parts supplier community. After an extended period of inactivity, Edelbrock was perhaps the stimulus of the renaissance of the Ford FE when they released the Performer RPM cylinder head. Scat followed a few years later with the release of stroker crankshafts, and then others followed with a steady stream of new and improved FE parts. Today, we can choose from hydraulic roller cams from Comp and Crane, blocks and heads from Pond, BBM, and, of course, Survival, and a huge array of cosmetic and performance items from every major brand-name supplier.

One last note is something of a dedication to my father, Seymour Rabotnick. He could not turn a wrench or identify a single engine part. But he understood my enthusiasm and gave me a set of tools and an old junk car to get me started along this path.

Building, or rebuilding, an engine is a significant task to take on under any circumstances. There are plenty of great reasons to do it, ranging anywhere between financial necessity and the personal satisfaction of completing a complex and challenging job. Once you've made the commitment to build your engine, a broad range of possibilities (economic, cosmetic, and performance) must be considered.

This book is targeted toward the more basic engine-building project. It could be a completely stock rebuild for a cruiser or daily driver, a restoration-style build for a muscle car, or a comparatively mild hot rod engine build. I anticipate the reader of this book to be working in his or her home shop or garage and to contract out the more sophisticated machine work. As your projects get more complex, or as performance needs grow, I suggest acquiring the Max Performance series of books for additional information.

At points in the building process you will be handing off the parts to a machine shop to cover work that you either cannot or should not try to handle at home. Discussions of the process with your chosen shop are important, and you will want to establish a good working relationship with them. How much of the work you have them take care of is predicated upon your budget and desired outcome. If you choose to do some of the work yourself using home-style techniques, just be cognizant of the potential impacts to your results. Throughout this book I list certain processes or products as "need to do," "nice to do," or "do if you wish." Sometimes saving a few bucks simply is not worth it.

This book is dedicated and directed toward the rebuilding of the Ford FE series of engines, but many of the procedures and concepts can be applied to any engine-building project. Throughout the book I try to point out and compare some options in terms of relative need versus cost or benefit. Often, several different acceptable methods can lead to a successful end result. I usually use a couple simple euphemisms to describe and evaluate choices for the task at hand. The guys at my shop hear the following sayings frequently.

The first one is "good enough really is good enough." I'll use this

This book will illustrate the processes for rebuilding your FE engine, but at some point, you will need the help of a machine shop or engine builder to complete the task. While not a requirement, it is always best to try to find a shop that has experience in building the same type of engine that you are working on.

phrase to indicate that certain parts, processes, or methods may be inarguably better than other choices while not being at all necessary for a given engine build. You can view this as being a "good, better, or best" selection process, or as insurance, or as an investment toward future plans.

Another frequently used comment I make is that a particular part represents "a solution looking for a problem." In the world of marketing there will always be folks trying to differentiate their product by making claims of better cost, performance, or longevity. If you read claims of huge benefits from a rather mundane and well-established product, it's usually a good bet that they are overstating their case to make the sale.

The last one I'll mention here is my use of the concept of "risk versus reward" as applied to engine building. Risk can be viewed as excessive cost, missed opportunity, or inadequate parts for the intended use. The "reward" has to be viewed in the context of the engine we are working on. In the racing world, engines have been refined to the point that any incremental gains are coming in very small increments. A $1,200 piston ring set that delivers 4 hp more to a professional race team has no real significance to a guy freshening up the engine for his pickup truck.

FE Engine History

The following is largely taken from the Max Performance FE book published a few years ago, and serves as a useful introduction to the history of the Ford FE engine. Throughout these pages you will occasionally find a comment or photo from that book, which makes sense, given they share the same author and publisher.

The FE Ford engine was released into production in 1958. The earliest applications included use in the short-lived Edsel program. The FE was not a replacement for the Y block; it was a larger companion engine family sharing some design features. In 1958, the Y block was still considered a current design at only four years old.

Starting out at 330 ci, the FE quickly grew in displacement through its first five years of production, with 352-, 390-, and 406-inch variants followed by the now-famed 427 in 1963. By 1966, the release of the 428 and the short-lived 410 completed the variety of basic passenger car combinations. A lot of high-performance history was written in a very short time. The 352 and 410 were dropped after 1966, and the 390 and 428 continued as the only FE engines in passenger car production from 1968 through 1970.

The FE had been dropped from passenger car use by 1971, but the 360 and 390 versions remained extremely popular in pickup trucks through the 1976 model year. Some commercial applications, notably U-Haul trucks, had FE power through the 1978 model year. Throughout the 20-year production run, the FE saw use as a marine, commercial, and industrial engine as well.

While the high-performance factory engines were responsible for all the glory, most production was for more mundane applications. Certainly the most popular original FE vehicles were full-size family cars and pickup trucks; they serve as the source for most of the engine blocks we start with today.

The beginning of the FE performance program had its roots as Ford split the car lines during the late 1950s, going from one basic plat-form to many as the market developed. The emergence of the bigger cars coincided with a gain in popularity of motor racing. The NHRA U.S. Nationals were held at Detroit Dragway in 1959 and 1960, and auto executives were exposed to the rising popularity of the sport. At the same time, NASCAR began the transformation that would take it from the local circuit group to national popularity. Television was about to change the way cars were marketed, and motorsports was one of the beneficiaries.

Ford responded to the market opportunity with high-performance iterations of the 352, then the 390. This was still the era where a production-based engine could be equally successful in both drag racing and NASCAR.

The FE performance program started out as upgrades to passenger car engines, using strategies that had been employed by hot rodders for several years. Higher compression, multiple carburetion, and dual exhaust were initially enough to get attention. But as the rivalry between the Big Three heated up, they quickly evolved into performance-specific engines. The first of these was the 406, blessed with a larger bore than the 390, solid lifter cams, and optional multiple carbs. Within a couple years the 406 was replaced by the 427, with a still larger bore, cross-bolted main caps, and better cylinder heads. The 427 became the lead piece for all of Ford's big-block race development and remained in that position through the end of direct factory involvement in 1970. When discussing professional racing and FE engines, you are almost always going to be talking about the 427.

The 428 was originally released in 1966 as a torque-oriented cruiser

Ford achieved incredible success with its FE-powered racing program, winning repeatedly at Le Mans, on the NASCAR circuit, and in the drag racing arena with their 427-powered Ford Thunderbolts, as well as the 1968½ 428 Cobra Jet Mustangs.

engine, but in the late 1960s somebody at Ford finally realized that the low-production, high-strung 427 was not reaching the masses. Ford had a good race program, but was getting a bad street rep. The more mundane 390-powered cars could not keep up with the big-blocks from GM or Chrysler. The response was to blend the readily available and bigger 428 block with higher-performance parts, including heads, cam, and intake. The 428 Cobra Jet package was available from late 1968 until 1970 and delivered on all points; it was reliable, strong, and still a competitive combination in NHRA-class racing.

The 429-engine family was slated to be the replacement for the FE, but the factory programs surrounding the new engine were short lived, barely making it two years before performance development stopped. Eventually the potential for the "385" family engine was realized, but that is another book.

The Famous Cars

Ford's initial platform for FE performance and racing was full-sized cars, the most popular being the higher-end Galaxie. Many FE engines were installed in full-size cars, most of them 352s and 390s. But the racers got the 427 cars.

While the 427-powered Galaxie was a good-looking and competitive package, it became quickly apparent that the Chrysler cadre had a distinct weight advantage with their smaller cars. The first response was to develop a lightweight factory drag race version of the 427-powered Galaxie. It included a high-riser version of the 427 engine, along with a variety of weight reduction strategies, including changes to sheet metal, interior parts, and even the frame. Always rare, and quite valuable today, the lightweights were only the opening act.

The next step was a factory authorized dedicated drag race car: the Fairlane Thunderbolt. The T-Bolts were assembled at Dearborn Steel Tubing, a Ford contractor. It took the lighter-weight midsized 1964 Fairlane sedan and installed the high-riser 427 engines into about a hundred of them. This was never intended as a street vehicle, and everything was modified to enhance the car's chances at the drag strip. Ford included major front-end work to accommodate the large engine, lightweight seats, thin glass, alumi-

num and fiberglass components, and race-only rear suspension. The Thunderbolt became a Ford racing icon, and the combination remains near the top of NHRA Super Stock racing 54 years later.

Ford did not install the 427 in a production Fairlane until 1966. The production 427 Fairlanes from 1966 and 1967 were rare, very cool cars with a solid racing history. But, like the lightweight Galaxie that preceded them, they never received the adulation reserved for the Thunderbolt.

Something about the almost absurd combination of small car and huge engine makes anything else seem normal in comparison. The ultimate expression of small car/ huge engine is also FE-powered: the 427 Cobra. The Cobra started out as the well-documented combination of a British sports car and a Ford small-block V-8 for road racing. Competing with well-funded efforts from both domestic and foreign racers, the need for more power was satisfied by grabbing an existing race engine, the 427 FE. What had already been an attractive sports car morphed into a beauty born of necessity, with broadened and flared fenders for larger tires, side exhausts, and a scooped hood. Brutal in both potential and execution, another automotive icon was born. Today there are many, many more inspired iterations of the car than were ever originally made. The 427 Cobra was and is the automotive definition of "badass."

Carroll Shelby first plucked the 428 Police Interceptor engines off the assembly line for use in the Shelby GT500 Mustangs in 1967. In 1968, the PI engine was replaced mid-year by the 428 Cobra Jet variant, and the car became known as the GT500KR. The 428 CJ continued through to the end of the Shelby production run in 1970. By 1969, the Shelbys were slow sellers, many unsold 1969 models were retagged as 1970 models. The cars are considerably more popular today.

NASCAR racing was the primary development test bed for Ford's FE race program throughout the 1960s. The 427 was upgraded and altered every year as needed to remain competitive. But while NASCAR served as the engine technology source, the cars themselves were not inspiration for many production performance offerings. Street enthusiasts looked to NASCAR for entertainment, but to the drags for inspiration. So, while we'll use parts that were designed for the high banks, we don't often emulate the cars themselves. Street cars have the big tires on the rear, scoops on the hood, but no numbers on the doors, a tradition that holds true today.

Throughout the late 1960s, professional drag race programs evolved, and the cars got further from a production basis. The hard-core drag racers moved into AFX cars, with radical modifications to wheelbases and engines. These in turn evolved into Funny Cars with tube chassis and nitromethane. The SOHC FE engine remained a common powerplant in these, but far removed from the engines available at the local dealer. These cars and engines are certainly worthy of discussion, impressive by any measure, but outside the context of this book.

The most famous of the FE-powered cars were never really sold to the public. Ford made a very public, concerted effort to get an outright win in the 24 Hours of Le Mans race in the mid-1960s. The first cars they produced were powered by the small-block engine. In subsequent years, the need for more power became apparent. In a situation similar to that of the Cobra, Ford looked to the already well-developed 427 FE as a power upgrade to the GT racing program. And the engine delivered, powering the winning cars in 1966 and 1967.

So here we have the FE engine legacy. The engine that was in the most famed Ford racing vehicles of the time in each form of motorsports, NASCAR, the Cobra, the GT40, and the Thunderbolt. This should be the backdrop for comparable fame and dominance on the streets of America. But it never happened. What went wrong?

The Normal Cars: Mustangs, Galaxies, Fairlanes, and Trucks

As a dedicated Ford fan and a Detroit-area FE racer since the 1970s, it hurts to say this, but it needs to be said. What went wrong is that Ford put everything into the low volume racing efforts and comparatively little into the everyday cars that made up the volume of production.

The FE was factory installed or available in numerous car and truck platforms. The full-sized Galaxie (and sister models) was the recipient of most FE production, from the early 1960s right up to the end. Most popular among enthusiasts are the 1963–1967 models.

Ford's intermediate cars, the Fairlane, Torino, and Mercury variants from 1966 through 1969, also had the FE as a regular production option. Most by far were 390-powered. A very few 1966 and 1967 models had a 427, and the 428 CJ was available beginning in late 1968.

Mustangs and Cougars were often FE-equipped from 1967 through 1970. The 1967 and 1968 big-block models were all 390-equipped. In 1969, there were a few 390s, but the 428 CJ was the engine of choice. The hydraulic lifter version of 427 was installed in a few Cougar GTEs in 1968 (replaced by the 428 CJ midyear), but no 427 Mustang has ever been documented despite 40 years of rumors.

Ford pickup trucks carried the FE as an available option through 1976. There are probably more FE engines in pickups than in any of the cars. The FE can be installed into any of the cars or trucks where it was an option. Any deserving small-block or 6-cylinder-powered candidate can be upgraded to FE power using factory replacement components

When new, a 390-powered Galaxie of 1964 or earlier was a competitive car on the streets and local tracks. But by the 1970s, it was common knowledge that the average 396-powered Chevelle could pretty much hammer any 390 car at will. A 428 Mustang could hold its own, but most FE owners simply lost enthusiasm. They got tired of getting their butts kicked every Friday night. They moved on to other cars or other hobbies, and the cars were left to sit or used as basic transportation. Interest from the aftermarket never really took off, so the supply of new parts was not there, and the old factory parts were getting used up and worn out.

By the 1980s, the FE engine was considered obsolete by all but a few die-hard enthusiasts and racers. No mainstream magazine coverage, no new aftermarket parts, and no real development outside the private efforts of the dedicated NHRA Super Stock and Stock Eliminator racers. The engine design that had won Daytona, Le Mans, and the Winternationals was considered obsolete and in the same league as the Buick Nailhead, the Chevy 409, the Olds Rocket, and Ford's MEL and Y block.

The Dinosaur Reawakens

But there was a difference: the cars. The Cobra was still worshiped, the Thunderbolt was still an icon, and the legacy from those early NASCAR, Le Mans, and drag racing wins still hung on. Stock and Super Stock racers running FE power continued to win with no factory support. As people started to repair, reproduce, and emulate those cars, the demand for FE parts began to build.

Specialty suppliers, including Dove, carried the FE flame through the slow years, catering to the dedicated racers and restorers. But when Edelbrock released a replacement FE aluminum cylinder head in the mid-1990s, demand finally began to build. A lot of candidate engines

While the 427-ci FE engine was all about dancing around redline on the tachometer, Ford knew that just wasn't very appropriate for some of the chassis needing upgraded big-block power. For the 1966 models, Ford introduced the 428-ci engine, which was a long stroke version of the FE engine design, specializing in comfortable torque rather than high-revving horsepower. Ford used the new 428 in full-sized Fords, Mercurys, and Thunderbirds as an upgrade to the 390.

came from the huge truck population. And there were a lot of candidate cars to choose from.

In 2004, Scat released a cast stroker crank for the FE, and Genesis concurrently released the first cast-iron reproduction 427 blocks. I built one of the first big-inch FE engines that used both parts, topping the 505-ci package with an EFI system. The engine was covered in *Hot Rod* magazine's July 2004 issue as the "676-Horsepower Dinosaur."

I entered a similar 505-ci FE in the Jegs Engine Masters Challenge the following year, using the new Blue Thunder cylinder heads. Most of the competitors thought it was pretty cool to see one of them ol' FE motors in the contest, and at first viewed it as a curiosity. Only after it made 752 hp on pump gas was it apparent that this was not a nostalgia piece; it was a modern engine with FE architecture. We finished 8th overall out of 50 entrants and got another magazine article as a result.

Jay Brown out of Minnesota entered his FE-powered 1969 Mach 1 into *Hot Rod*'s Drag Week competition in 2005. This is a grueling event covering more than 1,000 miles and five drag strips over a five-day period. The best overall average ET wins, and the Mach 1 took home the class win. Brown recently repeated the feat in a SOHC-powered 1964 Galaxie.

Subsequent FE race wins, engine builds, and project cars have gotten an increasing amount of media coverage from writers looking for something different. With a full array of parts now available, it is possible to build a complete 427 FE from scratch using all aftermarket pieces. You can build a 445-ci 390-based FE stroker that'll get you 500 honest horsepower without breaking the budget. In a few short years, the FE engine has gone from near extinction to mainstream again. This is without question the best time in the 50-year history of the FE to build one for the street.

PLANNING THE ENGINE BUILD

One of the first things to define is your anticipated budget for any engine building project. It is very, very easy to get overly excited about things when you start out, and go far beyond expected costs. The volume of unfinished project cars offered for sale should give you a clear idea of what happens if things get out of control.

Budgeting Process

A well-thought-out budget process involves several aspects. Some of these are rational, some are emotional, and all are important to consider before you grab that first wrench.

The cost and value of the vehicle and engine should be one consideration. If you are restoring a 1969 Shelby GT500, you can obviously justify investing a lot more into an engine than if you are building a scruffy 1976 F250 as a retro shop hauler. If that Shelby engine has a partial VIN stamping on it, you will want to salvage that block no matter how bad it may be. The worn-out 360 in the pickup has essentially zero market value and can be readily replaced if it needs significant repair. If you're

When deciding to rebuild an FE engine, many factors figure into the approach. Is it a garden-variety 360 truck engine? No need to go to any great length to save original components. If it is an original restoration candidate like this one, more thought must go into originality and value.

assembling a hot rod from scratch, you can set your budget in dollars and work backward from there.

The "risk versus reward" discussion is going to enter into the budget talk as well. When building a 300- to 400-hp engine, we do not need to consider the more exotic and expensive parts. The cheapest parts you can find are rarely (if ever) the right answer, but many common upgrades are fairly inexpensive in the context of a build's eventual cost. Just remember that each decision usually spirals into the next one, and that you will likely be spending more than you anticipated in almost every case as things come

together. Allow yourself some cushion on the financial side of things.

As you go through the budgeting process, you should also consider future plans and the ease of subsequent upgrades. If your project is bouncing up against the limits of your checkbook (as mine always do) think about the future. It's far easier to swap out for nicer valve covers, intake, carb, or distributor after the engine has been installed and in service for a while than it is to move to forged pistons or change cam types.

Throughout the actual rebuild process, you will be faced with decisions about whether to reuse and rework the original parts or replace them with new items. Some of these are normal wear items and are assumed to be intuitive; you are going to acquire new piston rings, engine bearings, gaskets, timing set, and such. Some items are potentially reusable but should be replaced in the context of a true rebuild. In this book I am assuming a minimum of new pistons, new cam and lifters, and the complete machining and reconditioning of heads and block. On the cylinder heads in particular, consider the cost of reworking the originals against the price of new replacements.

As a quick reference, at the time of writing this book the cost of stock replacement–type parts and complete machining will easily approach $2,000. Even when doing all the disassembly, inspection, and assembly work yourself, I would keep a $3,000 or $4,000 minimum budget in mind for a proper rebuild. It's very easy to double that (or more) when contracting some of the work or adding in upgraded or higher-performance parts.

Performance Goals

Performance goals have a direct relationship to the budget process. Almost everything you do to improve performance will have an impact on the project's cost. With that fact noted, quite a few performance improvements are possible for a modest cost increase. On the risk versus reward scale we can get a nice initial increase with very low risk/cost.

The first question when looking at performance goals will be the intended use of the engine. It might be fun to ask for 600 hp, 20 mpg, a smooth idle, and a low budget, but reality dictates that you are not going to get all of those at the same time. The right answer for a 4-speed Mustang is different from the proper package for a four-wheel-drive truck. It's most common to ask for a certain amount of power and then try to fit that engine into your design and budget envelope. It is often a better idea to approach it from the opposite direction: Define the vehicle needs and budget first, then see what kind of power you can get within those boundaries.

From the factory, the non-high-performance Ford FE engines used in passenger cars and light trucks were good running, durable, and reliable. But they were not noteworthy for outright power. They were intentionally designed for low-RPM torque; smooth, responsive driving; good idle quality; tolerance for low octane fuels; and low maintenance. In today's environment we have much better ignitions, more consistent (perhaps not better) fuel quality, and enthusiast owners who are much more involved in terms of tuning and more tolerant of high-performance characteristics such as idle quality, noise, and part throttle behavior.

A well-built stock or moderately upgraded 390 4-barrel engine should provide between 300 and 400 hp. A similar 428-based engine should deliver between 350 and 450 hp. While the 352 and 360 engines are worthy powerplants for street use, the reality is that you are far better off converting them to 390 cubes or more during the rebuild process. The upside gains in power and torque are dramatic, and the costs are nominal.

Limitations that should be considered included fuel tolerance. A compression ratio between 9.5:1 and 10.0:1 will usually work well with pump premium. Heavy vehicles with highway gearing should trend to the low side of that range, while a lightweight vehicle with steeper gears can go to or beyond the high side. This has more to do with the amount of load the engine sees going through the torque peak than with the absolute numbers. You also need to consider vacuum at idle if you are using power brakes. The bigger cams and larger carburetors associated with high power will reduce available vacuum, possibly below the minimum 10 or 11 inches desired at idle.

In this book we will restrict discussions to street performance applications within a range between 300 and 450 hp. We will primarily be covering builds based on 360, 390, and 428 engines. You can certainly make a *lot* more power using aftermarket or factory high-performance parts, and the comparatively rare 427 block, but that type of build falls outside the context of this volume.

Formulas for Engine Design

A group of formulas and equations are employed in both stock and high-performance engine building. Some of these are used to determine the basic configuration of the engine, such as displacement or compression ratio. Others are used during assembly to verify measurements such as deck clearance or ring gaps.

These days most of the calculations are readily available on the Internet or as part of engine-building software packages. Instead of listing them all out in complete mathematical detail, I will describe the purpose of the most popular ones, give a simple overview of the math and the impact they have on the build, and provide a couple examples. I will cover the measurement and clearance numbers during the assembly chapters.

Displacement

This is the measurement of "how big" an engine is in terms of cylinder volume. Displacement is referenced in cubic inches (or cubic centimeters). It is a simple cylinder-volume calculation, multiplied by the number of cylinders in the engine. Neither the combustion chamber nor the piston shapes affect displacement.

To calculate this number we need only the diameter of the cylinder and the stroke of the crankshaft (the distance the piston moves up and down during each crankshaft rotation).

To calculate displacement, we take the cylinder bore's radius squared (the bore diameter divided by 2 then multiplied by itself); multiply that value by PI (22 divided by 7); and then multiply the result by the stroke. You now have the displacement of one cylinder. An increase in bore diameter or stroke gives you more displacement.

This formula can be simplified and calculated as: bore x bore x stroke x 6.2832 inches.

An example for a .030 over 390 would be:

```
        4.080 bore
      x 4.080 bore
      16.6464

      16.6464
      x 3.780 stroke
      62.9233

      62.9233
      x 6.2832
      395.36 cubic inches
```

Compression Ratio

This is a comparison of the total volume in a single cylinder when the piston is at the bottom of its stroke versus when it is at the top. The combustion chamber and the pis-

ton have a great deal to do with this more complex calculation.

Compression ratio is expressed as a value over 1. We are comparing the total volume above the piston when it is at the bottom of its travel to the total volume above the piston when it is at the top of its travel. If we have 10 times more total volume at the bottom of the stroke than we do at the top of the stroke, we have an engine with a 10:1 compression ratio.

To perform this calculation we need the bore diameter, the combustion chamber volume, and the head gasket volume. You'll also need the deck clearance volume (the distance from the top of the piston to the top of the block when that piston is at the uppermost end of its stroke travel). The last thing we need is the effective dome volume of the piston; add this if it's a dish or subtract it if you have a dome.

Take the individual cylinder volume number calculated in the displacement discussion. Add in all the head gasket, combustion chamber, crevice, deck clearance (volume of that small space calculated as a cylinder), and dome volumes (some of those are usually given in cubic centimeters, which you'll need to convert to cubic inches). The total number is "on top" of your ratio. Now take all those volumes except for the displacement and you have the bottom number in the ratio. On a street engine you should end up somewhere between 9 and 10 to 1. Higher compression ratios will deliver more power, but will not tolerate low-octane pump fuels. On the risk versus reward scale, an extra point in compression might get you another 30 hp but cost you the need for race gas or a detonation-prone combination.

The typical calculation for a normal 390 Ford with flat top pistons (rounded numbers) is:

Bore = 4.050

Stroke = 3.780

Cylinder volume =
  (calculates to 48.71 ci)

Chamber volume =
  72 cc (converts to 4.39 ci)

Deck clearance =
  .030 (calculates to 0.39 ci)

Gasket volume =
  10.2 cc (converts to 0.62 ci)

Piston dish volume =
  6 cc (converts to 0.37 ci)

Total volume with the piston at the
  bottom of its stroke = 54.48 ci

Total volume with the piston at the
  top of its stroke = 5.77 ci

Compression ratio =
  54.48 divided by 5.77 = 9.44:1

Deck Clearance

This clearance is the result of a stack up of component dimensions. It is the distance between the top of the piston at its uppermost travel and the head gasket deck surface of the block. These days we seem to prefer to get this as close to zero as we can without going positive. The currently accepted ideal range for best power and combustion is for the piston to be between .040 and .060 away from the cylinder head at top dead center. To achieve this with the common .040-thick head gasket, we need to have the piston somewhere between .000 and .020 below the block's deck.

Add up one half of the crankshaft stroke, the center to center length of the connecting rod, and the compression distance of the piston. Compression distance is the measurement from the centerline of the piston's pin hole to the upper flat surface of the piston.

Example (using a 390 Ford FE and rounding up for simplicity):

Block deck height from factory measured from main center to deck surface = 10.17

1/2 of the 3.78 stroke = 1.89
Connecting rod center to center 6.49
Piston compression distance 1.76
Total = 10.14

Block deck height minus the total of parts = .030 deck clearance

This means that a simple clean-up machining of .010 to the block's deck will get you into the desired range.

Piston compression distance is a different dimension, and is determined when the piston is manufactured. This is measured from piston pin centerline to the top flat portion of the piston's head. This may not be the highest point on a piston; a domed piston will have a portion protruding above this point. Your piston manufacturer will usually provide this dimension for you in the instructions or packaging. It can be difficult to accurately measure on your own.

Ford FE Specific Design Choices

With some of the basic concepts in the background, we can take a look at the stuff that makes an FE Ford engine build unique. This will involve a bit of history to provide context to the possibilities within the factory architecture and using factory parts. While several aftermarket stroker packages are available now, this book is focused more on rebuilding and upgrading popular factory-style engines.

The engine pictured here is a completely original 1969 390GT. It has on it about 11,000 miles. The owner installed the vintage Cal Custom valve covers shortly after purchasing the car new, and it had been in storage for decades before he decided to refresh it and put it back into service.

Despite changes over the years, the majority of parts will physically interchange from one FE engine to another. It's a bit unusual to be referring to engine changes as "new versus old" when talking about an event that occurred 50 years ago, but that's what we do. The camshaft thrust design was altered in the early 1960s. You don't see many of them around, but the old engines can be easily converted to the newer configuration. The motor mounts changed in 1965; old ones used two bolts while the newer configuration has four bolts per side. The new-style blocks can be mounted into the old platform but the older ones need some adaptation to work in a post-1965 vehicle.

The most common Ford FE engines by far are the 352, 360, and 390. Used in hundreds of thousands of passenger cars and trucks, these are the engines you will most likely find in cars, barns, and salvage yards. Less likely although still possible are 391 and 361 medium-duty truck engines. It's also possible to find 428 blocks since they were used in full-sized cars and Thunderbirds, as well as in irrigation and industrial applications. The odds are very much against finding a 427 or 406 engine anywhere outside of the performance and restoration marketplace.

The acquisition cost of those engines pushes them to the outside of this book's budget-oriented focus, but all the concepts and processes described still apply.

The 352 starts out life with a 4.00-inch bore and a 3.50-inch stroke. The 360 starts out with a 4.05-inch bore and the same 3.50-inch stroke. A 390 has a 4.05-inch bore and a 3.78-inch stroke. With that noted, I tend to turn any 352 or 360 into a 390 almost by default. The cost of a 390 crankshaft and rods is very low, and the gains from the additional 30 or 40 ci are significant. A true winner in that risk versus reward equation.

The 428 is a special case with a 4.13-inch bore and a 3.98-inch stroke. Despite what you may see on the Internet it is best to assume that you *cannot* overbore a 390 block to 428 dimensions. These are thin-wall castings, and odds are you will eventually split a cylinder even if you get it running. More 428 crankshafts seem to be available than there are blocks, likely a testament to the crank's durability when other parts failed in racing efforts. The 390 block and 428 crankshaft combination will yield a 410-inch engine, a factory package used in a very few Mercury cars for a year or so. If you already own a 428 crank this is worth doing, but

The engine we are primarily concentrating on rebuilding for this book is a cool 428 CJ out of a 1969 Shelby GT500. The processes and machining are the same as we would find in rebuilding a more common 390 for a pickup truck or Galaxie, or any FE engine for that matter. I use other engine images where necessary to illustrate a particular process or concept.

piston availability is limited and often pretty expensive, putting your build cost into the realm of a stroker kit.

Build Process

I am going to make some assumptions here. The first is that you intend to build and assemble the engine yourself; hence the purchase of this book. The next is that, as part of this task, you will take the engine components to a machine shop at certain points in the build to get things done that are not possible in the average home shop.

Every engine-building project requires decision-making over a wide range of options. One very important series of decisions revolves around selecting the level of build quality. For some folks the decision is entirely based upon economics, and any way to save on costs is key. For others it's always going to be around quality, with a focus on having and doing things in the best possible way regardless of expense. Throughout this book I present a few options in both product and labor that range from "must do or have" to "a nice thing to consider." I assume the reader is not building a professional-level race engine and not working with his (or her) own fully equipped machine shop.

The lowest-cost option for a chosen process frequently will deliver a perfectly good result, simply trading time for expense. Other times, nothing but professional tools and talent will deliver the needed outcomes. Some tasks can be bypassed with minimal risk for a budget-oriented build, and others simply must be addressed in every effort for any realistic chance of success.

I use a single engine for most pictures and build process documentation throughout the book, and follow it along as we go. While this engine is a rather "cool" one (a 1969 428 Cobra Jet destined for a Shelby GT500), the processes and machining are exactly the same as found in rebuilding a more common 390 for a pickup truck or Galaxie. Most of the photos in this book are following the rebuild of the 428 Cobra Jet, but other engine images are sometimes used to illustrate various components or processes. Some photos were "posed" for better visibility, and may not reflect normal or proper machine shop practices.

We will follow this engine through teardown, inspection, cleaning, machining, and reassembly. During the teardown and inspection phases we will be "hands on" until handoff to the machine shop. At that point we will become spectators, watching the

shop handle its tasks. I do not detail how to run an SV-10 Sunnen cylinder hone, but I do explain what it is doing and why. Once the machined parts are completed, we will then resume our firsthand position as the builder doing the actual measurement and assembly work.

Choosing and Qualification of Your Core

This is when the fun begins. By definition, a "core" is the engine you start out with. It may be complete and running or a collection of parts gathered over time. Acquiring a non-running engine with an unknown history is like buying a lottery ticket, but you can tip the odds in your direction a bit.

My favorite type of core is old, greasy, and unmolested. Something still mounted in the vehicle is almost always more desirable than one that has been laying out open to the elements. When working with stock or nearly stock engines, the fewer indications of modification or prior internal work the better off you are likely to be in terms of internal condition. The 390 in a rotted-out pickup with an aftermarket Holley carb and some glass packs has likely had the snot run out of it, while the 2-barrel unit in an old LTD probably ran smoothly until the car fell apart around it.

Blocks cast in 1964 and earlier have a two-bolt motor mount. Blocks cast later have a four-bolt motor mount. The later blocks can be mounted easily into an older vehicle, but putting an old casting into the later application will require creativity and fabrication.

Look for obvious signs of distress. A prior owner trying to diagnose an engine problem will pull one valve cover and/or the oil pan chasing a knock. He may have removed a few spark plugs looking for coolant, or he could have pulled the distributor looking for a twisted-off oil pump driveshaft. Pushed-out core plugs or coolant in the oil might be signs of freeze damage in northern states. Fresh gaskets on timing covers or heads are a giveaway of recent repair work. A coat of inexpensive paint on an otherwise unremarkable engine might be as simple as a cosmetic sales pitch, or a tip-off to recent fix-up attempts.

Just remember to keep your eyes wide open and realize that the odds are very, very much against finding any sort of super deal on a 427 or 428. Pretty much everything you find will be a 360 or a 390, and those two are nearly impossible to identify externally. If you are able to turn the engine over with a socket on the damper bolt, you can use a wooden dowel stuck through the spark plug hole to identify the stroke by marking it at the top and bottom of the piston's travel. If it has about 3.5 inches from top to bottom, you're looking at a 352 or a 360. If you get a reading of around 3.75 inches you're onto a 390.

Block Markings

A good place to start your identification search, and a way to eliminate certain possibilities, is with the casting marks on the block. FE engine blocks usually have a number of casting numbers, both formal and sand scratches, on various areas of the block. Some of these marks are good for identification, but unfortunately many other markings were used almost at random and have little if any meaning for actual identification. I cover the most common ones below, but remember that nothing on an FE is to be taken for granted. We've seen actual non-cross-bolted 427 industrial engines as well as paper-thin 390s sold as standard bore 428s online. Take *nothing* for granted.

Mirror 105: Just like it says, a backward, mirror-image number "105" casting mark commonly found on the driver-side front face of blocks cast at Ford's MCC foundry starting somewhere in the early to mid-1970s. Usually a later-model 390 block with the extra main webbing. But not always.

352: The 352 designation is found on the driver-side front face of many of the FE blocks cast at Ford's DIF foundry throughout the 1960s. This does not mean you have a 352 engine. Or anything else for that

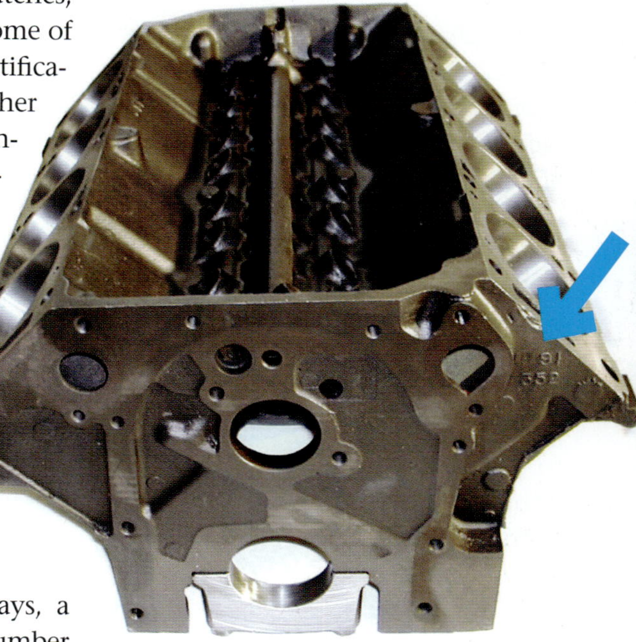

The 352 designation is found on the driver-side front face of many of the FE blocks cast at Ford's DIF foundry throughout the 1960s. This does not mean you have a 352 engine. Many other displacements have the 352 designation cast into the block.

The "DIF" casting often found on Ford FE blocks designates the Dearborn Iron Foundry where the blocks were poured. This location was in use through the early 1970s; therefore, a real 428 CJ block would likely have that DIF on it somewhere.

Similar to the "352" often found cast on the front face of the engine, the 352 designation in the bellhousing face does not guarantee you have a 352-ci engine.

This 427 marking that is often found in the lifter valley or the bellhousing face shown here is misleading. It can often be found on 390 engines as well.

matter because most 390 and 428 engines as well as many 427s will have this marking.

66-427: This one is often found on the inner valley above the lifters or on the bellhousing face. It tends to get folks really excited for a few minutes, but means pretty much nothing. Often found on otherwise normal 390 engines.

C scratch: This is a good one to find. Found as a freehand letter "C" scratched in the bellhousing area of the block, this is considered a good indicator of the 1968 and later double-webbed 428 block as used in the 428 Cobra Jet engines.

A scratch: Another nice find. This is the letter "A" scratched freehand into the bellhousing-area casting. Normally associated with 1966–1967 non-CJ 428 engines.

Inside the water jackets: Proof positive of a 428. If you remove the center freeze plug you can often see the number "428" cast right into the base of the water jacket core. Similar casting identification can also be found by looking straight down through the water opening on the decks where the head gaskets go. You'll need a flashlight.

Casting numbers such as C6MA-xx: These numbers are normally found cast upside down below the oil filter mounting pad. Unfortunately they don't really mean all that much. While important for a restoration project, the fact is that Ford used the same casting number across a wide variety of engine sizes and levels. That means that these numbers do not help for identification other than for exclusion. You know that a D4TE (the "D4" indicates 1974 in Ford code) is not going to be a 352, which was stopped in 1966.

Date codes: Often, but not always, cast in place above the oil filter pad, the date codes tell you when the block was made. Like the casting number, these will not tell you anything about the engine itself other than by exclusion (e.g., a block cast in 1964 is not a Cobra Jet since those started in 1968). Date codes are the holy grail for restoration work, but have limited value for performance efforts.

Cross bolts: Probably a 427, unless they've been added by a racer somewhere in the block's history.

Screw-in freeze plugs: Probably a 427, unless they've been added by a racer somewhere in the block's history.

The Drill Bit Test

This one test is the single best way to quickly identify an assembled FE block. Credit for it goes to FE.com forum member David "Shoe" Schouweiler. You need only the simplest of measuring tools: drill bits. The following is paraphrased from several of Dave's responses to block ID questions posed on the forum.

Remove the center freeze plug from the side of the engine block. Using common drill bits, try to slip the shank portion of the largest possible bit between the center cylinder cores through the freeze plug opening. The size of this largest drill bit will indicate which water jacket core was used to cast the block.

If you can fit only an 8/64-inch or 9/64-inch drill bit shank between the cylinders at the largest gap position on the block, and a 10/64-inch doesn't fit anywhere, then they are 427 water jackets.

The 406/428/DIF361/DIF391 blocks will allow a 13/64-inch drill bit shank to fit into the gap at the largest position.

The MCC361FT/MCC391FT blocks (MCC = "mirror 105" marking) allow a 14/64-inch bit to fit between the cores.

Regular 360/390/410 blocks have about a 17/64-inch to 19/64-inch water jacket space at the largest position on the block.

These are approximations, but they tend to be close.

Even if you do have the good jackets, be sure to sonic map the cylinders before boring because core shift might cause problems. It is not at all unusual for FE engines to have considerable core shift, and the oft-raced and abused 427 engines seem to have some of the thinnest cylinders.

Teardown and Inspection

Once you bring your new jewel home (and/or remove your "old friend" from the car) the real work begins. You can learn a huge amount from the teardown process. The key is to avoid the temptation to fire up the impact wrenches and rip it apart as quickly as possible. Careful inspection of an old engine will carry a clear history of the conditions it ran with, and will avoid unnecessary expenses if problem areas are identified and quantified early on. I am going on the assumption that we are taking the engine completely apart before

Be prepared for the occasional surprise when tearing down an engine with an unknown history. This 427 engine is a true barn find, and was filled with . . . mouse stuff.

handing the major components over to a machine shop for reconditioning. Be prepared for the occasional surprise.

In this chapter, I go through the basic teardown effort; details on each key component appear in the chapters that follow. We are going to work from the outside in as we go, removing external parts first. As you proceed, it's a great idea to use a digital camera to record how things came apart, and use plenty of plastic sandwich bags to label and organize the fasteners and small parts as you go. When parts are oily it's difficult to write on them; tag wire and paper tags are often useful. When labeling a box or a part for its location, I prefer to use the terms "driver-side" and "passenger-side" because they are less likely to be confused. In my shop we do almost nothing but FE engines; thus, we are quite comfortable with using just the Ford OEM cylinder numbering system with the cylinders on the passenger's side as numbers one through four. However, in a shop that works on numerous engines from many manufacturers, a description of "driver-side second cylinder" could be less likely to cause confusion.

You will want a reasonably large area to work in, with plenty of workbench space to lay out the parts as they are removed. It's a grimy process, so be prepared with lots of paper towels and rags. I find that covering the floor and work areas with newspaper or butcher paper goes a long way in controlling the inevitable oily mess.

Basic mechanic's tools are obviously required. If you are working with rusty fasteners you should be using six-point sockets to minimize the risk of rounding off the bolt heads. An air or electric impact wrench can be really handy for some stuff, especially the damper bolt. But try not to get too fixated on using it for the more easily removed stuff; the risk of damaging the bolts is simply too high to justify saving the 10 minutes of hand removal. Some of the external fasteners will be easily broken with an impact wrench, particularly the exhaust manifold bolts, which have seen many high-temperature cycles and exposure to the elements.

Start out by removing the carburetor (drain any gas) and the fuel pump. Both are likely to be rebuilt or replaced unless new or particularly valuable. Remove the spark plugs, plug wires, and any wiring harness segments that remained attached to the engine. If still in place, remove the alternator and power steering pump and their respective brackets. Pictures of these before removal will come in handy during assembly. Remove the pulleys, the oil filter mount, and the motor mounts.

Begin Teardown of Peripherals

1 *To remove the distributor, first take off the hold-down bolt and clamp. Then grab the distributor body, rotate it back and forth a couple times to loosen it, and pull straight upward. If it appears frozen or does not want to come upward, it is most likely held by dried oil and varnish around an O-ring that passes through the intake manifold, not by anything mechanical. Use some solvent such as carburetor cleaner and let it soak in while working the distributor back and forth to free it up. A few moderate taps on the bottom of the body with a plastic mallet are okay, but resist the temptation to pound on the vacuum advance. You will break the fragile distributor casting if you do and end up needing a new distributor.*

2 *Next remove the fuel pump. The fuel pump is attached to the timing cover with two fasteners.*

3 To remove exhaust manifolds use six-pointed sockets and wrenches, as these fasteners are often both tight and very corroded.

4 It is highly recommended that you use a generous amount of penetrating oil and sharply smack the wrench to break them loose. If and when they break off, use drills, bolt removing tools, and Heli-Coils to re-create threads in these locations.

5 When smacking the ratchet assembly does not work, try a ratchet with a long enough handle to provide greater leverage on the bolt head. Since manifold bolts often snap loose suddenly, visualize where your knuckles will go when they do. If the path they will take involves something metal and hard, reposition yourself.

Remove Rocker Assembly

1 On Ford FE engines you need to remove the valve covers and rocker assemblies before you can remove the intake manifold. This is different from any other V-8 engines. Take off the valve covers and invert them on the workbench. They make handy "trays" for the valvetrain parts you are about to remove. If you are trying to keep things in their original places, you can use a Sharpie or paint pen to mark them "driver's" and "passenger's" side. Punch holes in a piece of cardboard to hold pushrods in marked order if you intend to reuse them.

2 The rocker arm assemblies are each held in place by four 3/8 bolts, with 9/16 wrench heads and large, thick washers. Loosen each fastener only a half turn at a time, working from bolt to bolt. This will prevent bending the rocker shafts from valve spring pressure. Each rocker assembly will have a sheet-metal oil return tray underneath it that also gets pulled off with the rockers. Pushrods simply pull out at this point; they go through holes or passages machined and/or cast into the intake manifold.

Important!

3 On most FE engines one of the four rocker bolts is longer than the others, usually with a reduced diameter in the shank. This bolt is where the oil feed for the rocker arm system is located; make certain to note its position for reassembly.

4 With the rocker assembly removed, you can see more clearly the difference between the oil feed for the rocker assembly, shown second from left, and the other three conventional mounting points.

Remove the Intake Manifold

1 It's easier to use a knife to cut the water pump hose apart at this stage, as this is a replacement item.

2 If you have the factory iron intake be aware that these things are heavy at more than 85 pounds. If you still have the engine hoist available, it is very useful for this task. Remove the 10 fasteners holding the manifold down, taking notes that the rearmost ones are either shorter or have a spacer sleeve on them.

3 I find it helpful to use a razor knife to slice through the front and rear seals before trying to pry the manifold loose. An assortment of large pry bars, a very large chisel, and a plastic-covered dead blow hammer may be needed to get the intake to break free if it's been on for a long time. Be strategic and use common sense here. You do not want to damage the sealing surfaces or break the casting. Once you get one side or corner to come free, you can work it up and down to get the other side loose.

4 It's tempting to pry up the lifter valley tray and pull out the lifters at this point. But you stand a good chance of bending the sheet-metal tray up if you do. Better to wait until the heads and their respective gaskets are removed. Our engine was missing the valley tray. It is not easy to use with roller lifters, but a recommended item with flat tappet systems.

Remove the Cylinder Heads

1 With that intake finally out of the way, you can remove the cylinder heads. They are held down by 10 1/2-inch-diameter head bolts; 5 long ones under the valve cover area and 5 shorter ones alongside the exhaust ports. While I prefer hand tools, this is a place where a properly handled impact wrench with a six-point socket is okay for removal.

2 You will probably need to pry and wiggle the heads up a good bit to break them free from the old gaskets. To keep them from popping loose and dropping to the floor, I will often keep a long head bolt loosely installed in the center "short bolt" location as a precaution.

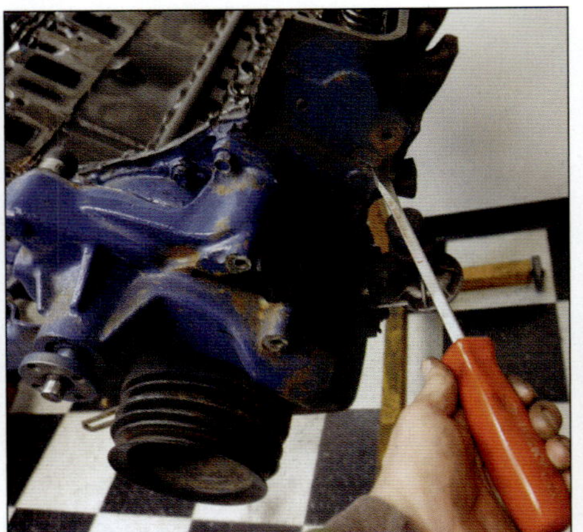

3 Be very careful with where you insert any pry device needed. We have seen heads where the gasket sealing surface has been nearly destroyed by heavy-handed removal damage.

4 A few light taps with a dead blow hammer can help coax the liberation of heads from block.

5 With the cylinder heads removed you can now pull off the head gaskets and easily pull the valley tray off. Lifters should theoretically just slip out with your fingers, but often take some effort on older engines. I have found that a spray and soak with carburetor cleaner or solvent helps loosen grime and varnish built up around the lifter's bottoms. A strong magnet, a small screwdriver for gentle prying, and a pair of pliers to grab the top might come in handy on lifters that don't want to slip out easily.

6 While most builds will be using new lifters, a very gentle touch will be needed if you plan on saving them for reuse (not advised), and keep them in exact position order.

Remove Water Pump and Front Damper Assembly

1 Next remove the water pump. The water pump is mounted to the block with four 3/8-16 fasteners. Pay attention to where you remove them from, since some may be unique for attaching accessory brackets.

2 The back of the water pump has a removable flat cover plate. If you are re-using the pump, change that gasket out.

3 Next is to remove the damper from the front of the crankshaft. You will probably need an impact wrench to loosen the bolt. It has a 15/16-inch hex head, and it is going to be very tight. Alternately, you can use a large breaker bar and figure out a way to keep the crankshaft from rotating (a large friend holding the flywheel or a fabricated contraption to stop a piston in its bore).

4 *With the bolt out of the way, use a dedicated puller to remove the damper. These tools are inexpensive and available from numerous tool suppliers, or they can be rented at many local parts stores.*

5 *It takes three bolts to grab the damper and a single long bolt in the center with a free spinning center that bears against the crankshaft. As you tighten up the center bolt it will pull the press fit damper off. I have seen folks use an impact wrench on the puller, but it is probably best to do it by hand to avoid damaging the tool.*

6 *With the damper out of the way, you can see the damper spacer and the square key. These are the next items we will remove.*

7 *Gently pry the square key out with a screwdriver. These usually offer very little resistance, but they are easy to lose. Be sure to keep yours in a baggie with the damper spacer. They are inexpensive and readily replaced if damaged.*

8 *The damper spacer is just a sleeve for the front seal to ride on and is not physically fastened in place. But they can get tight and bound in place over time. If it won't pull off you can try tapping around it with a plastic or brass mallet. In some cases, we have had to use a 2-inch muffler clamp and a three-jaw gear puller to get them loose.*

Remove Timing Cover and Timing Chain Assembly

1 Removing the cast-aluminum timing cover is a straightforward matter of removing the bolts. Most of the fasteners are 5/16 inch and have a 1/2-inch hex head, except for the two lower ones in the front on each side, which are 3/8 inch and require a 9/16-inch wrench. Keep track of which bolts go there.

2 There are also four fasteners on the oil pan that thread into the timing cover. These must be removed before attempting to pull the cover away from the block. With all fasteners removed the cover should pop right off with only a few taps from a plastic mallet and minimal prying.

3 After the timing cover assembly is removed, store it with the various sized bolts that go with it for easier reassembly.

4 Now that you have the timing cover removed, you can see the timing chain and gears for the first time. Depending on the reason for this engine needing a rebuild, it is always interesting to note how much free play is in the timing chain before you replace it. Careful inspection during disassembly can help resolve many mysteries.

5 *Next pull the oil slinger, which is just a piece of stamped sheet metal, off of the crankshaft snout.*

6 *Go to the single center bolt in the cam sprocket and remove it. It has a 5/8-inch head and might be tight enough to justify the use of an impact wrench. Be certain to make note of the position of the thick washer behind the cam bolt, and the fuel pump eccentric.*

7 *With the cam sprocket center bolt and washer removed, you can usually get the timing sprockets and chain off with some wiggling, prying from two sides at the same time, and perhaps a light tap or two with a plastic mallet. Brute force is rarely, if ever, needed; the trick is to keep both sprockets parallel with one another and even with the block as you go back and forth until the one on the cam comes loose. Then you can remove the chain and cam sprocket and concentrate on the crankshaft sprocket, which may be stuck to the keyway. If you find a large flat split washer as a spacer behind the cam sprocket, discard it now. None of the replacement timing sets use them, and trying to reinstall one with a new timing set will cause serious interference problems. Since the timing set is almost always going to be replaced, you can use a chisel between the teeth to split or spread that crankshaft sprocket if needed, but it should slide off with a bit of effort.*

Remove the Oil Pan and Pump

1 *If you're working on an engine stand, it's now time to spin it over and remove the oil pan. The pan is retained by lots of 5/16-inch bolts. Make sure you get all of them out before trying to pry the pan loose.*

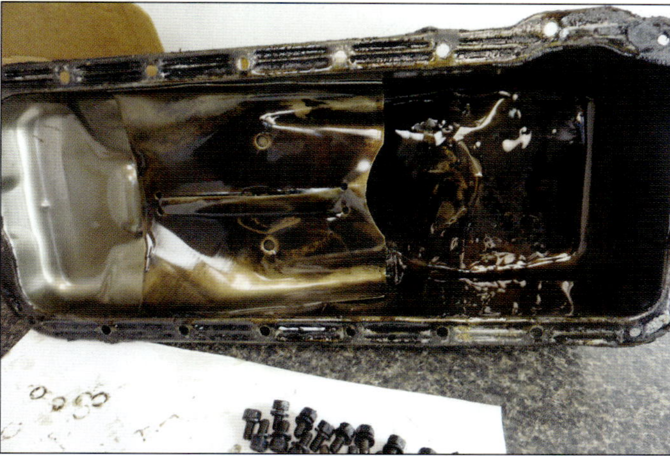

3 Badly damaged oil pans should be replaced; they are readily available. This one has seen better days.

2 The pan gasket will have become pretty well stuck to the block after time in service. I often use a razor blade or knife to walk around the sides and try to break that seal a bit before prying on the easily bent sheet-metal pan itself.

4 If your engine has a windage tray it will be sandwiched between the pan and the block with an oil pan gasket on each side.

5 You must separate the windage tray from the block using a scraper to split the gasket free.

6 With the pan removed, it's a quick task to remove the oil pump pickup screen and the pump itself, two bolts each. Then remove the oil pump drive; it will lift up with your fingers. This engine shows a lot of debris from deteriorated and fractured nylon teeth from an old timing chain set.

7 On everything beyond the lowest-cost effort we highly recommend replacing those items, but hang on to them for now.

8 Set the windage tray aside for later cleaning and possible reuse. Aftermarket trays are available, but original reproductions are no longer made.

Remove Piston and Rod Assemblies

1 Piston and rod assemblies come out next. The rods and caps should be marked for position before removal. When replacing both the pistons and rings during the rebuild, it's not so critical to mark those items, but it's a good working habit to get into.

2 A worn-out engine is likely to have a significant ridge around the top of each cylinder at the high point of piston ring travel. A ridge reamer tool is available for removing this ridge to ease removal and prevent damage to the pistons during a low-budget rebuild. But these days it is very rare to see such an approach taken. Instead, we simply assume that a ridge indicates the need for new pistons and knock them out, accepting the fact that they are not going to be reused.

3 *Most stock FE Ford connecting rods are nut and bolt types with floating pins. Working one assembly at a time, we remove them, working front to back. Rotate the crankshaft until the big end of the rod is low in the block so the fasteners are easily reached. I usually loosen both nuts partially, leaving them flush with the ends of the threads.*

4 *Next, I take a plastic mallet and give them a sharp smack, which will separate the rod cap from the rod while preventing the rod from falling out of position. With the cap now loose, you can remove the rod nuts and the cap, setting them aside on the workbench. Use boots or pieces of 3/8-inch fuel line hose to cover the bolt threads and keep them from hitting and marking up the crankshaft during removal. Using your hand to keep things in position, it's possible to rotate the crankshaft around to push the piston assembly partway up the cylinder. Continue rotating the crankshaft back out of the way and you can use a hammer handle or block of wood to continue pushing the assembly out of the block, guiding the rod to prevent cylinder damage. I use my knee or a helper's hand to keep the assembly from falling out onto the floor.*

5 *Once removed, you should lightly reinstall the matching cap and nuts to keep them together.*

Remove Crankshaft

1 *We are down to only a few parts left now. Remove the main cap bolts. An impact can really come in handy here, although a long extension handle will do the job. The main caps are going to be tight in the block. Most all FE engines have the main caps marked for position at the factory, but it is a very good idea to double check and mark them as needed for position and front to back orientation. After 40 years of service, it is possible that somebody machined them differently, or replaced one and messed up the original order.*

2 *You can use a combination of the loosened bolts and a mallet to wiggle the caps loose and up out of the block. The rear one can be really tight in there due to the rear seal assembly. A slide hammer can be attached to the oil pan bolt holes to serve as both a lever and a tool to add a bit of upward force to get it free.*

3 *With the main caps out of the way we can lift out the crankshaft. These are pretty heavy, around 70 pounds, so have a place cleared off to set it once it comes out. You may need to rotate it a bit to find a spot where the counterweights clear the block for easy removal. Grab the snout and the rear flange and pick it straight up.*

4 *Keep your main bearings and caps sorted for further inspection.*

5 *Now that the crankshaft is removed from the block, find a clean safe place to store it before bringing it to the machine shop for inspection.*

6 *Inspection of the old rod and main bearings can provide good clues about the condition of the crank and rods. Excessive wear, off-center wear, and circumferential scoring can help diagnose what is going to be wrong before we even begin to measure. This information can help your machinist pinpoint areas that may require extra attention.*

7 The old bearings can often provide useful information. These indicate that the engine was rebuilt in the 1970s, and that the crank journal has already been ground .010 undersized.

Remove Camshaft

1 To remove the cam you must first remove the front thrust plate. These are retained by two fasteners, which are often large Phillips-head screws. You must acquire and use the Phillips number-4 bit to remove these screws; anything else will round them out and create a bunch more work.

2 Once the thrust plate comes off, the cam will simply slide out through the front. By leaving this as the last major item to remove, we can guide it out with our hands since no other components are in the way, which is much easier than trying to pull it out the front without any support.

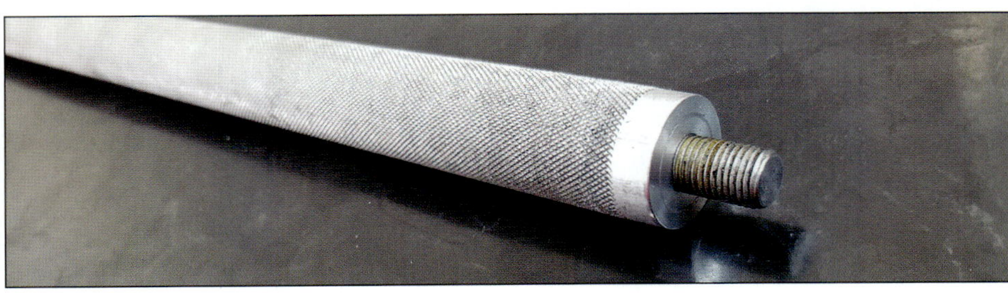

3 In our shop we made a cam removal tool using threaded rod and a long piece of aluminum. Even a simple long bolt with the proper thread pitch can be helpful here.

4 This handmade tool makes cam removal and installation much easier without the worry of marking up bearings or journals.

5 *Finally, before heading to the machine shop, we finish off a few details. The rear cam plug gets knocked out using a long stick (a cut-off piece of broomstick works amazingly well). Notice that the plug goes in backward compared to a freeze plug: very important. Freeze plugs can be removed by punching an off-center hole in them and leveraging them out with a big screwdriver or a slide hammer. That same slide hammer will make quick work of the press-in-type oil gallery plugs as well. Threaded-in gallery plugs can be removed with the appropriate tools, along with application of heat and lubricants. These can be really tough and sometimes end up requiring drills, taps, and foul language, but they do come out.*

6 *Cam bearings are normally removed by the machine shop, using the same type of tool used to install them. They are easy to knock out if you rent or borrow the right equipment. Do not try to remove them yourself with simple punches or you risk damaging the cam tunnel bore and scrapping the block.*

Now we are ready to take our parts to the machine shop for individual inspection, cleaning, and reconditioning as necessary or desired. Subsequent chapters focus on each component separately until we start putting things back together. Accordingly, you will find me doing some assembly work then backing up a bit to go over the next component in the process before moving ahead. I recommend that you read over the entire book before beginning the job to best familiarize yourself with the tasks ahead.

THE ENGINE BLOCK

We now have our engine block taken down to its bare essentials: a casting without any parts bolted to it. This is the point where we can decide if it is truly a good candidate to move forward with. A few fairly simple inspections and we are off to the machine shop, where things will get a lot more intensive.

As with other processes throughout the build, you can do this work in many ways, depending on your experience, the tools on hand, your budget, and your desired results. Plenty of engines have been successfully rebuilt with the most basic of tools, done completely in a home garage. Just do not expect such a low-budget package to compete with a professionally machined and reworked combination in terms of power or longevity. Keep your expectations in line with your abilities and your finances and you will be happy with the end result.

Engine Block Inspection

When we inspect a block before machining, we are usually looking for damage that cannot be repaired or cannot be repaired at a cost that can be justified. The cost versus value is very different depending on a number of factors.

Common issues include the need for cylinder sleeves due to excessive wear or physical damage. You may also find cracks from freezing or from prior component failure. Significant corrosion damage is possible in certain engines, such as raw water–cooled marine applications or castings that have sat outside for an extended time. Damaged bolt hole threads are common.

A common 360 or 390 block will justify only modest work before you reach its current $200 or $300 replacement value, which is maybe a single sleeve. A 427 block will almost always get repaired; their multi-thousand-dollar value can justify extensive work. If you are working with an original "numbers-matching" combination, you will do whatever it takes to retain that block because of the value it adds to the car.

Basic Block Dimensions

All Ford FE engines share many common characteristics. The nominal deck height as measured from the main bearing center to the cylinder head mounting surface is 10.17 inches. Blocks can be identified from there using a couple external cues, the casting numbers, the date codes, and the bore diameters.

Very early blocks have a different cam mounting and thrust control method. They can be converted to the later style with a little work, but

Basic bore diameter is a quick identifier of an engine block and can be measured with a dial caliper. Precision is not required for block identification, so there is no need to be concerned about to-the-thousands accuracy. A dial caliper is perfectly fine for this task.

A 428 begins life at 4.130-inch diameter. After 40 years its pretty safe to assume that if somebody is selling a supposedly standard-bore 428 with fresh machining, he is trying to push a .080-over 390 off on you. Stay away; the cylinder walls will often be tissue-paper thin.

A 428 block is often identifiable by a letter "A" or "C" that appears hand scratched in the bellhousing area.

A 428 block can also be identified by the numbers "428" cast into the floor of interior water jackets below the center freeze plug or below the front or rear water openings to the heads.

are usually not the best candidate for normal use unless you have all the parts and the need for earlier casting numbers.

Basic bore diameter is a quick identifier, and can be done with a dial caliper. We are seeking only general sizes here, so to-the-thousandths accuracy is not required at this point. A 352 engine starts out life with a 4.00-inch bore, whereas a 360 or 390 begins at 4.050 inches. If your engine block is under that 4.05-inch size, it's not a 390.

All the 360 and 390 blocks are between 4.050 if standard and 4.110 at .060 over. We generally try to avoid anything beyond 4.090 (.040 over). The sonic test detailed in a subsequent paragraph will demonstrate why. Most FE blocks being built will be either 390- or 360-based. We covered block identification in some detail in chapter 1, but some things merit repeating.

When checking casting codes for FE motors, casting numbers are more useful for telling you what you don't have versus what you do have. The first two characters in Ford casting numbers represent the decade and year; in the case of the FE, being primarily manufactured in the 1960s and 1970s, the letter "C" represents the 1960s, and "D" the 1970s following Ford's numbering system.

A 427 block is generally out of the context of this book, but with cross-bolt mains and a 4.23 bore they are easily identified. It is fairly common to see the numbers "66-427" cast into a common 390 block; don't get too excited if you see this.

Casting numbers are more useful for exclusion than for true identification. The first two characters in Ford casting numbers represent the decade and year in which the casting was released; C8 is 1968, D2 is 1972, etc. A "T" in the third position indicates that the item was initially designed for a truck application, while an "A" means it was originally designed for passenger car use.

Something with a D4TE-xx or a C4AE-xx number is probably not a 428 since those were made from 1966 through 1970, and never installed in trucks. There is often a date code cast below the oil filter pad that provides exact casting date information for the numbers-matching folks.

All FE engines, with very rare exceptions, use the same main bearing diameters, the same cylinder head bolt patterns, the same cam tunnel diameters, and similar bolt holes everywhere on the block. This means that almost every part can be transferred from one block to the other as long as you pay proper attention to things such as valve diameters against cylinder bores.

Example: C8 is 1968, D2 is 1972, etc. Something with a D2TE-xx or a C4AE-xx number is obviously not a 428 since those were made from 1966 through 1970. When it exists, the date code is cast below the oil filter pad. This code provides exact casting date information for the numbers-matching folks.

Fixes for the Factory Block

The factory FE Ford engine block is a very robust design, noted for its durability. Most damage to a factory FE block is related to normal wear or abuse/neglect and can be readily repaired in a well-equipped machine shop. Our initial steps with the block are focused on inspection, making certain that the block is usable and repairable. You want to minimize the possibility of investing a lot of money into a part that cannot be run.

Common damage includes cracks caused by freezing, stripped-out bolt holes, split or scored cylinder walls, and corrosion from sitting in an uncontrolled environment. Most of these issues are repairable within the limits of financial necessity, although a catastrophic failure will occasionally render a block worthless for most folks.

The first step in finding flaws and preparing for reworking the block is a thorough cleaning. Depending on your budget there are a few ways to get the job done. A really low-budget

There are three notable variations in main webbing. The aforementioned 427s with cross bolts fall outside my context here. But we still will often see single or double webbing in the main saddle areas on common blocks. The double-web blocks are stronger, and are used in 1968-and-later 428s, including the Cobra Jet, along with many later truck applications. Visual identification is easy once the pan is removed. Compare this example to the single-web block in the top image.

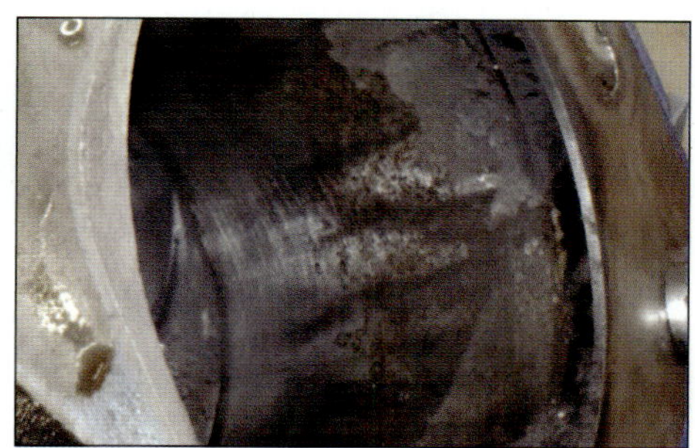

Common cylinder block damage includes cracks caused by freezing, stripped-out bolt holes, split or scored cylinder walls, and corrosion from sitting in an uncontrolled environment. The cylinder walls in this example are showing a fair amount of corrosion, which can be repaired by boring, honing, or in extreme examples, re-sleeving.

Once cleaned up, preferably through a professional process of heating the block to burn off gunk and corrosion, followed by an abrasive media blasting, many previously hidden flaws will be readily visible. Look for cracks alongside bolt holes and in the valley area and adjoining freeze plugs. Cleaning the block will reveal old repairs and hidden damage.

Shown here is a stripped-out head bolt hole. Although a serious problem, this can be readily repaired with a Heli-Coil.

One unique area to check on an FE engine is the oil feed passage that runs from the cam bearing up to the deck on each side. This hole feeds oil to the rocker arm assembly and has been known to crack on occasion, allowing oil to get into the coolant.

effort might be nothing more than a pressure wash and a lot of scrubbing. But to do the job real justice, we use a "shake and bake" process that involves heating the casting up to burn off any accumulated oil, grime, and paint, followed by an abrasive media in a tumbler that brings the block to a new-casting finish level.

Once cleaned up, many previously hidden flaws will be readily visible to the eye during a careful inspection. Look for cracks alongside bolt holes, in the valley area, and at adjoining freeze plugs. Block cracks can be further identified and evaluated with a magnetic tester, but you need to suspect where they are before you can really test them. Cast-iron cracks can often be repaired by either welding or pinning. Both of these are challenging repairs best put into the hands of folks with a fair amount of experience. It's not a really good place to practice a new skill unless you are willing to sacrifice a block to the learning curve.

Stripped or damaged bolt holes are common in old castings and are readily fixed with a thread insert. The most popular of the repair inserts is a Heli-Coil, where the hole is drilled and tapped to an oversize and a stainless-steel wire spring is wound into the hole. The result is a perfectly

good thread with no real downside. This work is well within the skill level of most any home engine builder, and with a little care can be easily done successfully. Making a simple drill jig will help you keep the holes straight and perpendicular to the part.

One unique area to check on an FE engine is the oil feed passage that runs from the cam bearing up to the deck on each side. This hole feeds oil to the rocker arm assembly, and has been known to crack on occasion, allowing oil to get into the coolant. You can pressure check this by covering the hole at the cam bearing end with your finger and putting just a bit of air into the top of the passage using a rubber tip air gun. It should appear airtight; if you hear air blow-

ing into the cooling jackets you have a leak. The fix is to install a sleeve (a piece of tubing) into the passage. While we make a sleeve in our shop, I have heard of folks using brake line tubing or old pushrods for the purpose. The sleeve needs to fit snug but not necessarily super tight, and be lightly tapped/pressed into place. We usually lightly coat the outside of the sleeve with silicone before installing it for added insurance.

Sonic Checking Charts

Our next step in preparation is to do a sonic test. This is also recommended when you bore a block out beyond .030 or .040 over because FE engines are noted for having fairly

Regardless of the type of build, it always provides peace of mind to do a sonic test, which is a test for the thickness of your cylinder walls. It is an optional process for most rebuilds, but a nice thing to consider for anybody working on a rare or expensive block. This is also recommended when you are boring a block out beyond .030 or .040 over, as FE engines are noted for having fairly thin cylinder walls.

Sonic testing is used to determine metal thicknesses in places that are not easily measured. To perform a sonic test you must first calibrate the tester to your block. The front face provides a spot that can be easily measured and the tester can be set up to match the actual measurement.

This is an example of a press-in gallery plug. It is cheap insurance to replace these with screw-in gallery plugs, as they are less prone to work loose over time.

thin cylinder walls. It's far better to know beforehand whether you should start off with a different block than to discover a thin wall or core shift after investing in the machine work.

A sonic tester uses sound wave reflection to give an accurate idea of material thickness. Testers can range from the very accurate commercial quality products to inexpensive online auction parts. This is a case where we are not trying for absolute accuracy, but rather for a good directional idea as to whether the block remains a good candidate. We will usually test each cylinder in multiple locations: at the top, the bottom, and center. On a common 360 or 390 block we are looking for minimum thicknesses of +/- .100 after machining.

Pre-Machining Block Preparation

We have our candidate. It's cleaned up and any critical flaws have been found and addressed. We are now ready to prepare it for machining. At my shop we follow a pre-machining regimen that falls under the "nice to do" banner. Some of these modifications are not really mandatory, but well worth the added effort.

We do this work before machining because we are working alongside crit-ical areas and we do not want to risk damaging a newly finished surface.

Oiling System Modifications

All the oil gallery holes are drilled and tapped for 1/4-inch NPT screw-in-type plugs. From the factory many of these are already screw in, but others use a press-in-type plug that can come loose after multiple rebuilds as the hole gets worn. The only hole we do not tap is the one that ends the driver-side lifter gallery up front, behind the distributor opening. That one gets really close to both the lifter and the distributor and is best left as a press-in plug.

Use a die grinder (you could use a Dremel or similar) and a carbide burr to open up the oil feed passage where the oil pump mounts. Most original blocks have a simple 5/16- or 3/8-inch hole there, which leaves a big step in the oil flow path. We blend it out to match the pump's outlet opening.

You can set a main bearing shell into each upper main position and scribe the oil hole position into the block. Original blocks are noted for having a portion of this feed covered up by the bearing. A few minutes with a grinder will line up the feed and improve flow. You are not trying

All the oil gallery holes are drilled and tapped for 1/4-inch NPT screw-in-type plugs. From the factory many of these are already screw-in, but others use a press-in-type plug that can come loose after multiple rebuilds as the hole gets worn. The screw-in plugs are installed using a bit of Teflon paste.

It's a good idea to open up the oil feed passage with a die grinder. This is often an area of restriction, depending on the block. Matching this opening to the outlet from your oil pump will ensure the best flow rate.

Opening the oil feeds in the main bearing shell will help oil flow, especially where bearing openings and the oil feed do not line up perfectly. Be careful here, only a small amount of material needs to be removed. Also check for cracks.

to remove a lot of metal here, just roll/blend over the last quarter-inch to line things up. A gentle touch goes a long way.

Casting Preparation

Run a clean-out tap into all the threaded holes on the block. Even after a thorough cleaning it's very common to find debris entrapped in the bottom of a blind hole. And having clean, straight threads makes assembly far more pleasant. Any damaged threads can be repaired using a thread insert such as a Heli-Coil.

Lifter bores get a quick dusting with a brake cylinder hone. We are not trying to remove any significant material, but many cleaning processes tend to roll edges and leave minuscule protrusions in hidden spots. The lifter bores are common victims since they require very tight and accurate clearances. They can be checked and repaired as required

during machining, but if you choose to save money and run them as-is, you will be really upset if a lifter won't go into place after assembling the rest of the engine.

Go over the external surfaces of a block removing excess casting flash and obvious flaws. I have never been enamored with the idea of polishing the lifter valley or detailing the external parts of the engine casting to perfection. You may choose to do that, but don't expect any gains beyond the satisfaction attached to the work itself.

Machine Work: The Big Stuff

We are now handing our prepped block to the machine shop for them to do their work. Three machining processes and one installation procedure in particular I always recommend having done at a professional shop with quality machining and measurement equipment. The machine work involves boring/honing the cylinders, milling the decks where the heads bolt on, and line honing the main bearing tunnel. The cam bearing work uses a set of installation tools not often found in the home shop, and if things don't go smoothly a machine shop is better equipped to address the issues found.

The order in which the processes are done will often depend on the shop, their equipment, and on the type of work required, so there is no need to dwell on that.

At least a couple of these procedures can be done at home with simpler tools if you are working against a really tight budget, but the end results will usually be far lower in terms of quality, durability, and performance. Given the amount of effort it takes to get an engine removed and torn down, the performance risk in

skipping the shop work is not normally worth the reward in terms of the money saved.

Cylinder Bore Preparation

The cylinder bores are perhaps the most critical part of a machined block. Although simple in appearance, they serve as the sealing surface for the piston rings and the bearing surface for the piston's skirts. They are also the part of the engine block that sees the most significant wear over time during normal use.

To do their job properly, cylinders need to be straight up and down, as perfectly round as possible, and have no noticeable scratches or damage. Engines can and do run with all manner of flaws, but any deviation from the target will have a cost in terms of power, durability, and oil control. What is acceptable for a farm or work truck is very different from what is expected of a performance build or a concourse restoration. Keeping expectations in line with investment is pretty important here.

The most inexpensive and home-builder friendly version of cylinder bore prep involves a heavy-duty drill motor and a flex hone of some sort. These hones can comprise either stones or balls embedded with abrasives. They are spring loaded and by design are really intended more for cleaning off debris and accumulated glaze as opposed to truly removing any significant amount of material from the bore. If you have a race engine getting a fresh set of rings, or a perfect block just receiving needed service, this can work quite well. If you have an older engine getting a complete rebuild, a flex hone will not suffice for any but the most budget-oriented project and will usually be marginal even for those.

As cylinders wear they tend to get larger in diameter at the upper part of the bore, and you will often see a ridge form where the top ring travels. This is because of the combustion

As cylinders wear they tend to get larger in diameter at the upper part of the bore, and you will often see a ridge form where the top ring travels. Part of the process to eliminate this ridge will be boring the cylinders to a larger diameter and fitting the appropriately sized new pistons and rings to match.

pressure pushing against the piston rings; it is highest at the top and gets lower as the piston is forced down the cylinder. When worn out, the rings are no longer able to effectively seal at the upper part of the bore due to the changed diameter, along with accumulated wear on the ring itself. To fix the problem we go to a slightly larger-diameter piston and ring package, and size the cylinder bores to fit.

To properly rework a worn cylinder bore you need to use a rigid hone, and possibly a boring machine, when going significantly oversize. Boring equipment is used to remove a large amount of material for oversizing or sleeving, or when you need to alter the original geometry or location of a cylinder (unusual in a mild build). A boring machine will use a single cutter to slowly go down the cylinder removing metal. The resulting hole may be very round, but is not at all suitable for running a piston or ring in. The rings absolutely require a far smoother honed finish be applied after the boring is completed, and a minimum of .003 inch of material must be left for finish honing subsequent to the boring operation.

Prepare the Cylinder Bore for Sleeving

1 *Shown is a block ready for sleeving. Sleeving is a valid repair for a damaged cylinder, or a cylinder that has already been bored beyond desired tolerances. This is a detailed process, and definitely not a do-it-yourself operation. If you look carefully, you can see the base ridge at the bottom of the cylinder, which is left from the original bore as a stop for the sleeve.*

2 *A sleeve can return the bore to its original size if desired, or it can be sized to match overbored adjacent cylinders, often depending on whether you are doing a single cylinder or multiple cylinders. It is best if all cylinders have the same bore since pistons and rings are usually sold in complete sets.*

3 *Although some shops use a big hammer and a block of wood to install a sleeve, I prefer to be a little more careful. William Blair, our resident machining guru, made a fixture that allows a sleeve to be smoothly pulled into position. The lower part of his fixture attaches to the main bores.*

4 The upper part of the custom sleeve-installation tool locates and holds the sleeve in position as it is smoothly pulled into the overbored cylinder. Doing it this way reduces the amount of press fit interference required, and is less likely to cause damage to an already injured block.

5 When the sleeve is pulled snugly against the machined stop left in the bore, a small amount will protrude from the deck surface. This will be perfectly leveled when the deck gets surfaced.

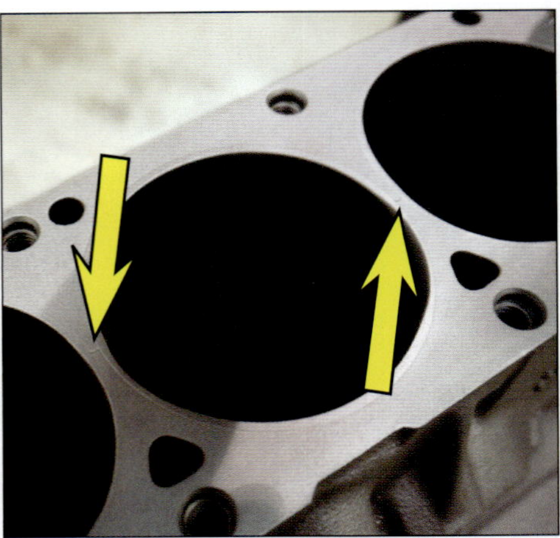

6 A sleeve installed with too much press fit, especially adjoining sleeves, can cause serious problems. The example shown here cracked in between each sleeve. It was not able to seal a head gasket as a result, and put water into the cylinders. This kind of damage is difficult, if not impossible, to repair in a cost-effective manner.

Sleeves are a valid repair for a severely damaged cylinder. They are a comparatively thin-wall ductile iron tube that is pressed into position in a bore that has been machined oversize. A sleeve can return a damaged bore to its original diameters if desired, or it can be sized to match other cylinders in the engine. It is a fairly complex item to properly install, and is decidedly not a do-it-yourself operation. It is important to have a step in the bore for the new sleeve to index onto as a "stop." Finances preclude the use of multiple sleeves. On a normal 360 or 390 a single sleeve makes sense, but multiples are not financially prudent

To properly rework a worn cylinder bore, you need to use a rigid hone, and possibly a boring machine when going significantly oversize. A rigid hone is used to establish the correct surface finish for rings, as well as to get to the proper final bore diameter. It can be operated by hand with a very large drill-style motor and a rig to support the weight, but most quality-oriented machine shops have long since gone to an automated honing machine. The computerized Sunnen SV-10 hone shown is state of the art for a well-equipped shop.

While not necessary in all applications, the use of torque plates is recommended. Torque plates simulate the stresses and deformation placed on the block as if cylinder heads were torqued into place, which is how your engine will spend the rest of its life. Almost universally used in racing builds, use of them in any rebuild provides greater accuracy and peace of mind.

Most machine shops will start out with a comparatively rough stone to get to within .0005 inch of the desired bore diameter. Then they'll follow with a smoother stone to reach the final diameter. These two steps are particularly important to create the desired plateau surface for the best results and the shortest break-in time.

Note the proper use of a torque plate, proper fasteners, and gaskets to simulate bore distortion caused by the torque of the head bolts. The cylinders shown were initially honed without a torque plate, and had low mileage. This image shows the amount of deformation that loads from the head bolts and gaskets create.

A tremendous amount of research and science has been applied to the correct hone finish for cylinder walls. Check with your piston ring provider for information on final bore finish, as well as your piston supplier for proper skirt clearance dimensions.

since the cost of the repair exceeds the value of the block. A 427 block can justify numerous sleeves because of its value. Adjoining sleeves can cause deck cracks to form in between cylinders, so your shop needs to be very careful during installation if you decide to follow that path.

A rigid hone is used to establish the correct surface finish for rings, as well as to get to the proper final bore diameter. It can be operated by hand with a very large drill-style motor and a rig to support the weight, but most quality-oriented machine shops have long since gone to an automated honing machine. While the hand operation can deliver good results in the hands of a skilled operator, the machine is both faster and far more repeatable for both accuracy and finish quality. With rare exception you will want to see a honing machine being used in almost any machine shop you select. Some honing equipment can be utilized, along with diamond abrasive stones, to do a modest amount of overbore work, speeding up the process and eliminating the need for a separate boring operation.

One optional addition to the honing operation would be the use of torque plates. These are flat plates that are bolted to the top of the block before honing, intended to simulate the stresses and deformation placed on the block by the tightly torqued head fasteners and the gaskets. Near universal in racing applications, the use of torque plates is a good addition to consider for even a stock rebuild because they provide better initial sealing and reduced break-in time (not at all mandatory as millions of engines have been built without their use, but I consider them a good return on the investment).

Current honing technique involves a multiple-step process that delivers a plateau finish on the cylinders. The process begins with a comparatively rough stone to get to within .0005 inch of the desired diameter. This is followed with a smoother stone to reach the final diameter, leaving a smooth and flat surface area for the rings to ride upon, along with a pattern of scratches that hold oil for lubrication and sealing. Some shops will follow up with an even smoother stone for a near-polished surface and brushes for cleaning the surface, but the initial two steps are particularly important for the best results and the shortest break-in time.

A tremendous amount of research and science has been applied to the correct hone finish for cylinder walls, including stone grit, depth and volume of the scratches, angle of the crisscross pattern, and the percentage of flat versus scratch voids. All of this is well beyond the scope of this book, and you will be entirely reliant upon your machine shop for the outcome.

A few things hold true no matter what process is employed. Always refer to your chosen piston ring supplier for general guidance on bore finish. Always be certain that the machinist allows at least .003 inch for honing stock to arrive at the final bore diameter. Always refer to your piston supplier for proper skirt clearance dimensions, as they determine actual finished bore diameter.

Although modern pistons are far more accurate in diameter than used to be the case, most machine shops will want the pistons to be on hand so they can accurately set the skirt clearance and thus the finished bore diameter. Too tight a clearance can result in scuffing the piston's skirts, while too loose will result in a noisy engine. The clearance is machined into the piston skirt, meaning that a piston intended for a 4.080-inch bore and .002-inch clearance will have a measured skirt diameter of 4.078 inches.

Decking

Another critical operation is decking, or resurfacing the portion of the block that the head gasket sits on. As in many other operations, you can go the low-cost route and simply clean all gasket material and residue from the deck thoroughly with solvent and sharp blades. After that, you can take a sharp, flat file and lightly wipe the deck to identify and remove any raised areas. This simple approach will work just fine in many cases, and relies on the gasket to handle any flaws.

But there are a lot of opportunities for deck deformation in a 40- or 50-year-old block, and a blown head gasket is not much fun if your budget approach fails. As a general rule, we deck every iron block that comes in to our shop; it's the best way to guarantee a good result. You do not need to remove much material, just enough to get a flat and smooth surface over the entire area.

It is very common for us to see raised areas around bolt holes, low areas between cylinders, and significant variances in deck heights and angles. Painting the deck with machinist's dye and making a very light cutter pass will highlight these areas, clearly illustrating why we do this work on every build.

Always square-deck your block. To do this, measure from the main bearing tunnel to the deck surface front and rear on both banks, find the lowest/shortest location, and machine both decks flat and to match each other, and exactly 90 degrees apart. On a race engine you will often machine to a target dimension to reach zero deck clearance, but on a more economical build you are really just trying to get everything clean, square, and equal.

Newer machines are able to

Resurfacing the portion of your block where the head gasket sits is called decking, and it is a very important step in the machining process. Decks often deform over time, and there is no more critical sealing surface on an assembled engine than the head gasket. Remove as little material as possible to achieve a flat deck and good gasket seal.

It is best to square deck the block. This is achieved by measuring from the main bearing tunnel to the deck surface front and rear on both banks, finding the lowest/ shortest location, and machining both decks flat and to match each other, and exactly 90 degrees apart. This ensures that everything is clean, square, and equal.

provide a very smooth deck surface finish. Older equipment was comparatively coarse and left a visibly grooved surface. The better finish quality is nice with any head gasket, but a really smooth surface has become mandatory if you choose to use the multi-layer-steel (MLS)–style head gaskets.

After decking and honing are complete, it is very important to add a very small chamfer or edge break at the top and bottom of each cylinder. This can be done with an abrasive cone, or a small, sharp half-round file, or a deburr tool. But a sharp edge left there will wreak havoc when you try to install piston and ring assemblies later on. Deburring all openings on a machined part is simply a good shop practice to follow.

Align Honing

For proper bearing fit, the main bearing bore needs to be perfectly round. All five bores need to be perfectly in line with each other to prevent crankshaft binding and

premature bearing wear. Over time, the bores tend to get deformed from loads applied to the crankshaft, becoming oval shaped and oversized. The align hone operation is intended to get all the main bearing bores perfectly round and in line with one another.

To make the bearing bores round and return them to original diameters, we must make them smaller again. This is done by removing a very small amount of material (a couple thousandths of an inch) from the faces of each main cap in a dedicated grinder.

The now shortened main caps are reassembled onto the block and the fasteners are torqued to specifications. A special long honing mandrel and stone assembly is pushed through the entire block in full strokes, bringing the diameter back up to the desired size and putting all the bores into alignment in a single operation. Done correctly, only a couple thousandths of an inch of material is removed from the block, having minimal effect on the cam tunnel–to–main tunnel dimensions.

Line boring is a different operation, and is done when major problems are present, main caps need to be changed, or significant modi-

fications are being made. Similar to boring a cylinder, it utilizes a single cutter to remove material. It has the benefit of being able to accurately correct alterations in main tunnel location, along with the risk of causing them so inadvertently. I prefer to run the align hone process after a boring operation in any case, and try not to bore anything that does not absolutely require it in mild builds.

Block Detailing

This is a repeat from notes found throughout the machining process descriptions. But they are worth repeating. We take the time to remove any sharp edges from alongside each freshly machined surface. It makes the block safer to handle, and it prevents any parts being installed from catching on a sharp edge and becoming damaged or raising a burr. I recommend chamfering the edges around each bolt hole and water jacket opening as well.

While we do not get overly fixated on polishing every internal surface, I will remove excess casting flash and edges that are protruding from the surfaces and parting lines where the casting cores joined up.

To return bearing bores to their original dimensions (and to make them perfectly round again), we first need to make the bores smaller so that we can bring them back out to OEM specs. This can be accomplished by removing a small amount of material from the faces of the main bearing caps. The shortened caps will provide a smaller bore that we can work with.

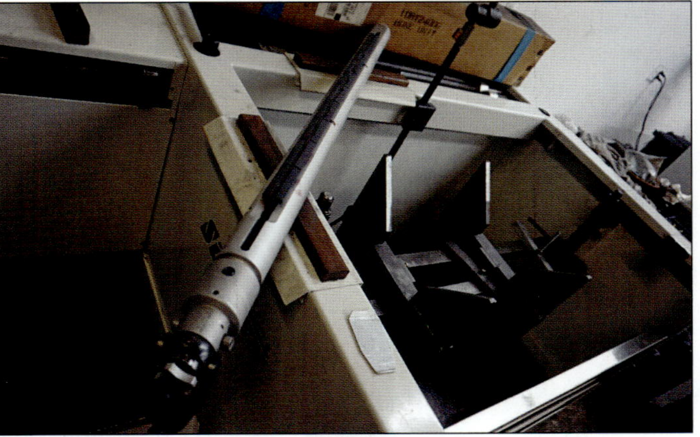

After the main bearing caps are torqued into place, this long honing mandrel and stone assembly is pushed through the entire block in full strokes, bringing the diameter back up to the desired size and putting all the bores into alignment in a single operation. Done correctly, only a couple thousandths of an inch of material is removed from the block to achieve the desired results.

We are now at the stage where we can get to cleaning up the block for storage before final assembly. Dirt, debris, abrasive materials, and tin whiskers will find their way into every nook and cranny imaginable. Use a pressure washer to clean the large areas and dedicated brushes with soap and water to get the rest.

Follow up by using an air gun to make certain that every possible bit of debris has been removed from all passages and surfaces. After air drying the assembly, use WD-40 on all machined surfaces to prevent rust from forming on your newly cleaned block while it rests.

A couple things are likely to trip you up if not checked and addressed at this stage of the project. Blocks used in trucks may require provisions for a rear sump–type oil pan. The oil pump pickup used with those pans has a support brace that needs to attach to either the block or one of the main cap bolts. If you need that type of pan it's important that your block already has the necessary holes or that you drill and tap them now. Some truck and industrial blocks will use a dipstick mounted to the oil pan, while passenger cars use a dipstick that presses into the block near the oil filter. If you are converting a truck block for passenger car use, make certain to knock the plug out of the dipstick hole. Last is that the medium-duty FT truck blocks have a larger-diameter distributor guide hole at the very bottom below the camshaft. If converting a medium-duty truck block for passenger car applications, you will need to use a bushing to reduce that opening to the normal size.

Cleaning is essential to a successful build and cannot be overstated. It seems that running and machining debris will find its way into every passage and hole. Scrubbing with hot water and soap, and use of pressure wash equipment before assembly even begins is pretty much a mandatory step in every single build no matter what the budget. After thoroughly scrubbed we will air dry the block and spray WD-40 on all machined surfaces.

Camshaft Bearing Installation

Cam bearings should be installed after a thorough block cleaning. Most FE engines will have an annular oil groove machined into the cam housing bores. Those grooves are impossible to clean once the bearings are in place, and any debris you miss will be entombed within the passages, just waiting to wreck your new engine.

The cam bearings are usually installed by the machine shop. FE cam bearings are different for each position in the block. They are all the same inside diameter, but the outside diameter is larger at the front, and becomes progressively smaller to the rear of the block.

Installation is done using a special tool, supporting each bearing as it is pressed or pounded into place. The commonly available "universal" installation tools use an expanding mandrel to hold the bearing, with a centering cone and shaft to keep things straight while tapping the bearing into position. Since we do so many FE installations at my shop, we've made a solid mandrel and press-in tool to better support the bearings. It's greatly improved

You will probably have the machine shop install your cam bearings. Most home DIYers don't do this job frequently enough to justify the investment in the special tools required.

Cam Bearing Diameters

Factory FE engines have different diameter cam bearings for each location in the block. While the inside diameter is the same fore to aft, the outside diameters get progressively smaller as you go toward the back of the block. If you do invest in the tools to do this yourself, pay attention to the diameters and make sure to align the bearing oil feed holes to the oil feeds. The front bearing has an opening that directs lube to the gear, so make certain that it is positioned correctly. On stepped cam bore blocks such as FE engines, it is easiest to install the rearmost cam bearing first and work your way forward.

The common 360, 390, and 428 blocks will have an annular groove behind the cam bearings in positions two, three, four, and five. This means that the orientation of the oil feed hole is not super critical in a street build. Racers prefer if the hole in the three o'clock position can be viewed from the front. On a rare 427 side oiler, you will need to orient the holes with the feed passages that lead up to the rocker arms.

Whether you have installed the cam bearings yourself or have had the machine shop install them for you, it is best to put motor oil on the bearings and test fit the cam you plan on using before beginning engine assembly. If you find bearing misalignment or other issues, you want to be able to fix it before all the other components are in the way.

Now that the block is clean and dry, it is an excellent time to throw a coat of high-temp engine paint on it. Not only is adhesion better when the block is this clean, the paint will help keep the block looking nice. I am not a fan of any sort of paint on the inside surfaces of a block, although plenty of folks do it.

our percentage of successful installations, but would be difficult to justify in a shop not doing a bunch of FEs.

Once you have installed all five bearings it is absolutely mandatory to test fit a camshaft into the block, preferably the one you intend to use. The camshaft should spin smoothly by hand without any tight spots. It is extremely common in FE engines for the cam bearings to install a bit out of line with each other, or "cocked" in the bore, making cam installation impossible. The time to find and correct any problem is now, not after you have all the pistons and crank-shaft installed. At this point any misalignment can usually be addressed with a couple of taps on the end of the installation tool to re-center the offending bearing, or some light scraping on the bearings, or even touching up the journals on the camshaft itself.

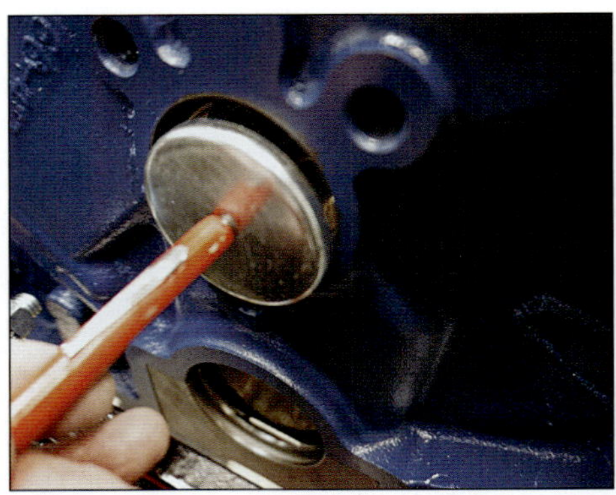

The camshaft plug is installed using a small amount of Motorcraft TA-31 silicone.

This is **important**: the cam plug on an FE has a reverse taper profile, and goes in backward in comparison to a normal freeze plug, with the flat surface facing out.

Block Final Work: Paint and Plugs

I usually paint the block at this stage. You will get the best paint adhesion while the block is dry and really clean, and it helps keep it nice. A simple coat of high-temperature engine paint is amazingly durable when applied to a clean surface, and can easily be painted over later in the final assembly stages for a show-quality appearance if desired. I never paint an interior engine surface such as the lifter valley, believing that the risk of the paint coming off far outweighs any incremental reward.

Oil up all the machined and unpainted surfaces with WD-40 or a similar product to prevent rust. A clean, machined surface will rust almost instantly.

Now you are finished with the basic block essentials. Put your finished block in a protective bag and let's move on to the next procedure.

Threaded oil gallery plugs are installed using Teflon paste. Press-in plugs get a dab of epoxy. Make double certain that you install the plug at the front end of the driver-side lifter gallery. It's an easy one to miss. If you have really low oil pressure on startup, this is the likely culprit.

The freeze plugs (properly referred to as "core plugs" if you're into details) on an FE are supposed to be 1⁴⁹⁄₆₄ inches in diameter. A lot of the inexpensive plug kits will include 1¾-inch plugs. Those are often too small and will not provide a tight enough press fit, causing eventual leaks or fallouts.

I normally use a small amount of that same TA-31 silicone for core plug installation, but they can be installed dry, or with epoxy if preferred. The most important thing is that they are the correct size, that the holes are clean and undamaged, and that they are installed correctly: straight and just below flush with the block surface.

CRANKSHAFTS

Original-equipment crankshafts were made from either cast-nodular iron or forged steel. The nodular-iron cranks have proven to be extremely durable and are found in the majority of production engines. A steel crankshaft is inherently superior, especially in severe service where exposed to extremes in terms of power or cycle fatigue. A road-race engine is exposed to extreme cycle fatigue because it is subjected to long periods of heavy throttle and transitional loading. A steel crank is ideally suited for this application. On the other hand, a cast-nodular-iron crank is perfectly adequate for the budget-constrained, street-oriented builds and many drag-race applications because these cranks are typically subjected to full-throttle operation for shorter periods of time. By far most FE builds will use a cast-iron crankshaft.

Steel crankshafts were installed in the always-rare 427 performance engines with their 3.780-inch stroke and in medium-duty truck 361 and 391 engines. The limited supply of factory 427 steel cranks in repairable condition drove their cost beyond the reach of many engine builders several years ago. There are a few NASCAR 427 steel cranks that use a wider connecting rod and bearing than common OEM equipment. With these supporting parts being nearly unobtainable, the NASCAR cranks are best left in the hands of collectors who have a demonstrated need for such parts.

The 3.780-inch-stroke steel 391 truck crankshaft can be converted into a performance piece. The truck crankshaft has a larger-diameter snout where the dampener mounts, as well as a different flywheel mounting flange and a counterweight combination designed for external balance. The pilot-bushing (or converter snout) hole in a steel truck crankshaft is larger in diameter as well, mandating a custom pilot or converter bushing for use in passenger car applications. The crankshaft nose needs to be cut down to the standard 1.375-inch FE diameter, the flywheel mounting surface is shortened by about .060 inch, and the counterweights need to be cut down before balancing is possible. The supply of usable 391 cores has also

FE Ford engines used a wide variety of crankshafts over the production life of the engine. All of them can be physically installed in any FE block. The most popular cast cranks were offered in strokes of 3.50 for the 352 and 360 engines; 3.78 for 390, 406, and 427 engines; and 3.98 stroke in the 410 and 428. Steel crankshafts were available only in the 3.78-stroke 427 engines for passenger car use, and in 3.50- and 3.78-stroke versions for medium-duty and heavy-duty trucks.

The steel 427 crankshaft pictured is both rare and expensive. Compared to the 427 unit, the steel truck crankshafts have completely different snout dimensions and counterweight designs. The easier-to-find truck cranks can be converted to passenger car and racing use by a skilled crank grinder. Before aftermarket crankshafts came onto the market, this was a common task for performance use.

dwindled in recent years, increasing the cost of the finished product, but this remains a viable option for those needing a steel crank within the stock stroke ranges at half the cost of a custom billet. The required modifications are time consuming, but a skilled crankshaft shop can handle them.

Identification

By far the most common factory offerings were the 3.50-inch-stroke cast crankshafts found in 352 and 360 engines and the 3.78-inch-stroke cast units common to 390s. The highly sought-after 3.98-inch-stroke cranks were used only in 428 applications and for two years (1966–1967)

Once you've seen an FE crankshaft, it's pretty easy to pick out another one among any collection of parts. They have a unique, long front snout similar to that of the 429/460 parts, but are much smaller in main journal and counterweight size.

This is a factory original 1969 390 crankshaft removed from an engine with 11,000 miles on it. You can clearly see the various factory paint markings as well as a balance hole in the rear counterweight.

in Mercury 410 engines. A few of the 428 crankshafts have different center counterweights and were specified for vehicles equipped with the Super Cobra Jet (the "Drag Pack" option).

Cast cranks such as this one can easily be identified by the thin-cast parting line running the length of the crank, as opposed to the thicker line of a steel crank.

Your odds of finding one of them in a normal swap meet or junkyard are slim.

It's a fair bet that the average FE crankshaft you find in a parts pile or swap booth will be a cast 390 or 360 piece. If you are really lucky, you may find a 428 crank. A steel truck crankshaft occasionally shows up, but most often they have been turned .030 or .040 under. A serviceable steel 427 crank is extremely difficult to find outside of Ford specialty retailers or among dedicated racers.

It's fairly easy to tell a cast crankshaft from a steel one. The cast crank has a very thin parting line where the two halves of the mold meet one another. In comparison, the steel

In comparison, this is a factory Ford steel crankshaft. You can clearly see the far thicker parting line, which is characteristic of a forged part. Any of the forged crankshafts will have this physical feature.

The forged steel 427 crankshafts often have the somewhat appropriate "$" on them. The original part number will often be partially ground off; that is normal for these and not an indication of modification.

crankshaft has a thicker parting line, usually between 1/4 and 1/2 inch wide.

Ford forges or casts numbers on the counterweights of the FE cranks, which may simplify identification. As is the case in most FE parts, the

This is a 428 crankshaft with the 3.98 stroke. All the 428 cranks (and only 428 cranks) will have a "1U, 1UA, or 1UB" stamping on one of the counterweights.

numbers are helpful but not definitive. Often they are useful for exclusion; if a crank is stamped "2U" it's obviously not a 428 part, but the absence of any stamping number does not preclude it from being one.

Here are some common casting marks; it's by no means a comprehensive list:

- The cast 360 often has a 2T or 2TA
- The cast 390 often has a 2U or 2UA
- The cast 428 has a 1U, 1UA, or 1UB
- The cast 428 SCJ has a 1UA or 1UB
- The steel 427 may have a somewhat appropriate "$" sign

These parts are now 30 to 40 years old and have often seen a vast amount of modification, alteration, repair, and service. By far the best way to identify the crank is to measure the actual stroke using V-blocks or even an engine block with a couple main bearings set in it. The stroke variances are large enough that you don't have to be perfectly accurate for basic ID; even a ruler does the job. It is around 3.5 inches for the 360, just more than 3.75 inches on the 390 cranks, and nearly 4 inches for the 428 version. The extra center counterweights on a 428 SCJ crank are obvious, as is the large

1.75-inch-diameter snout of an unaltered steel truck crank.

Measurement and Inspection

The most important descriptors for any crankshaft (OEM or aftermarket) are "solid," "straight," and "round." These terms apply to the crank as a whole as well as to each individual journal.

Checking the crank for straightness can be done either on a set of V-blocks or, lacking those, you can use the block itself by cradling the crank with bearing shells in the front and rear journals only. Use a dial indicator on the center journal to see whether it's straight or not. Try moving a bearing from one end to the center and rechecking at the end journal. While the factory-accepted tolerance per the book may be higher, anything beyond .0005 inch is a potential problem in an assembly where clearances less than .0030 inch are expected.

The same holds true for main and rod journals. These need to be checked in multiple places around each journal, and on at least a few spots front to rear. It is common to find a rod (or main) journal that has significant diameter differences

Check the journal diameters of the main and rod journals. These need to be checked in multiple places around each journal, and on at least a few spots front to rear. It is common to find a rod (or main) journal that has significant diameter differences forming either a tapered or barreled shape. Even new crankshafts can show variance in this area, and can be machined for uniformity.

When looking at potential crankshafts, take a quick look at the snout diameter, the condition of the keyway slot, and the threads for the damper bolt. While often repairable, damage in these areas often renders a crank financially unsound as a prospect. Compared to a passenger car shaft, a steel truck crankshaft snout is much larger in diameter, uses a much larger bolt, and would need modification to use in a car engine, which is an expense that falls outside the focus of this book.

forming either a tapered or barreled shape. Once again, any variance beyond a few ten thousandths of an inch can lead to early bearing wear or failure. Out-of-tolerance journals are common in used factory parts as well as brand-new items. You need to check. Your crankshaft machinist can grind to the next undersize if needed to straighten out a lot of these variances, a pretty common occurrence on even new imported crankshafts.

Additional dimensional specifications that must be considered include the rod journal width and main thrust journal width. Rod journal width, along with the dimensions of the chosen connecting rods, determines the rod side clearance. This is a more forgiving clearance with a broad acceptable range (.008 to .020 inch is common), but still something that needs to be validated during assembly. Thrust journal width can be compared to specs while actual thrust clearance, a critical dimension, cannot be determined until the crank is installed in the block with the chosen bearing.

Cranks should be checked for cracks using Magnaflux equipment. While a good idea for any part, this is particularly important when reusing a cast crank because it is very common to find fatigue cracks around the rod journal radii alongside the counterweights. Sometimes these are surface flaws that can be removed by grinding down to the next undersize, sometimes not.

Throw index and stroke variance are less often checked in the typical home engine build, but the latter, in particular, can be important to determine. If you're not paying attention, it can "catch you," and the consequences are catastrophic. A lot of folks target a near-zero deck clearance on their engine project, simply adding the nominal dimension of connecting rods, pistons, and crank stroke, and then machine the block decks to the combined value. Differences in stroke measured from journal to journal of .001 inch or more are not unusual,

and I've seen more than .005 inch on some import crankshafts. A bit too much will impact compression ratio, piston-to-valve clearance, and piston-to-deck clearance.

Before you pronounce your crankshaft candidate a winner, take a quick look at the snout diameter, the condition of the keyway slot, and the threads for the damper bolt. While often repairable, damage in these areas often renders a crank financially unsound as a prospect. With the cost of a usable or repairable 390 crankshaft being under a couple hundred dollars it does not make sense to spend too much on repairs.

Grinding and Polishing

Once you've decided to move ahead with your crankshaft, it's time to take it to a machinist who has the specialized equipment needed to recondition it. Many automotive machine shops will send their crankshaft work out to a specialist because of the cost of the equipment and the skills needed to do the job correctly.

The crank grinder is a large machine that uses abrasive wheels to remove material from the journals, returning them to a factory finish, but at a smaller diameter. On older engines such as the FE Ford, it is common to finish them to progressively smaller diameters in increments of .010 inch; hence, a crankshaft with both main journals and rod journals reduced by ten one thousandths of an inch will be referred to as "10-10."

After the journals are ground to size, they need to be polished to establish a good surface finish for the bearings to work with. The surface finish left by the grinding stones would be too rough and cause premature wear. The polishing is done with an abrasive belt, which removes very little material. After polishing, the crankshaft is ready to be thoroughly cleaned. Concentrate on running a brush through all of the oiling passages. Once it is cleaned, you can spray some preservative oil on it and put it into a bag to keep it safe until balancing and assembly.

Balancing

We are going to briefly cover balancing here, although it is normally one of the last steps before assembly. A big piece of balance work is focused upon the crankshaft. While balancing is not mandatory on a mild stock rebuild, it is usually a very good idea on any engine that is going to stray from factory weights or dimensions of parts.

Most FE Ford engines are internally balanced. This means that the flywheel and damper are balanced by themselves and have no effect on the balance of the engine itself. On 428 (and some medium-duty truck 391 engines) the balancing is external, requiring additional weight on the flywheel and sometimes the damper

On 428 (and some medium-duty truck 391) engines, the balancing is external, requiring additional weight on the flywheel and sometimes the damper spacer. Those engines will need the flywheel (and possibly the damper spacer) installed on the crankshaft for balancing. Note the size of the weight on this flywheel. A 428 crankshaft can be balanced internally to permit use of a normal 390 flywheel, but doing so requires heavy-metal (tungsten) slugs, which can double the cost of balancing.

Balancing is not mandatory for a stock-type build. But it is very important when new parts are significantly different (often much lighter weight) from the original parts. When balancing, a precisely measured weight assembly is clamped to each rod throw. These weights simulate the rotating and reciprocating weight of the piston and rod assemblies.

spacer. Those engines will need the flywheel and damper spacer installed on the crankshaft for balancing.

Balancing involves a couple of steps. We have "rotating weight" and "reciprocating weight." Rotating weight includes the large end of the connecting rods and the rod bearings. Reciprocating weight includes pistons, pins, rings, lock clips, and the small end portion of the connecting rods. All the weights used in balance calculations are given in grams: 454 grams equal 1 pound.

The first step is to weigh and match all the reciprocating components to one another. Pistons should weigh the same as each other, as should bearings, piston rings, and pis-

The balancing machine will spin the crankshaft and weight assembly and provide a digital readout. The readout will indicate the weight of material that needs to be removed (or added), along with the location of the removal.

ton pins. The small ends of the rods should also match and can be ground if needed to get them corrected.

The thrust bearing on an FE is in the number-3 position in the middle of the block. I have seen folks accustomed to building other engines put these in the wrong location. Place the upper main bearing shells (the ones with the oil holes) in position in the block. They snap into place dry with no lubricant or adhesive of any sort on the backside of the bearing.

The pistons in modern sets are usually very closely matched, while the other parts are near perfect. Pistons can have a small amount of material removed if necessary to equalize them, but this has become rarely necessary anymore.

With all the parts equalized, the weight values are input into a formula. The formula will provide a value that is referred to as the "bobweight." The counterweights designed into the crankshaft need to offset the bobweight to achieve a smooth-running engine. A selection of weights is mounted to the crankshaft rod journals to replicate the bobweight, and the crankshaft assembly is spun on a sophisticated measurement machine called a balancer.

We use the balancing machine to measure the location and amount of material that needs to be removed or added to the counterweights to achieve this goal. To reduce weight we will drill holes into the counterweights. To increase weight we will need to press in slugs of tungsten, a heavy metal that weighs twice as much as steel.

Bearings and Assembly

With a completed and *cleaned* block and a reconditioned crankshaft ready, we can begin the assembly process. Normally, I install the camshaft first as a convenience since it's easier to slide it into position before the crankshaft and connecting rods are in the way. We will be covering the camshaft selection in a subsequent chapter, so this is one of those sections where you will want to read

Bearing clearances are a critical measurement. The clearances published in most bearing catalogs are possible arithmetic ranges using parts that meet specification, and are not recommended targets. A good rule is to target .001-inch clearance per inch of journal diameter. A little bit tighter is safe on a street build, but you should always try to remain below .003 inch.

The most accurate way to check main bearing clearances is with the use of a dial bore gauge. To use the dial bore gauge, first install the bearings and torque the main caps to specification without the crankshaft installed. If you are building a 427 with cross bolts, torque those as well; they do change bearing clearances.

The bearings are snapped into the main caps the same way: clean, dry, and with no sealer or anything between the cap and the bearing. The bearings in this picture are Federal-Mogul race bearings. They have that dark, burnt appearance new, out of the box. They deleted the cosmetic tin plate operation to achieve a stronger bearing alloy.

ahead before starting. Make certain to use oil on the cam bearing journals and the appropriate break-in lubricant on the cam lobes.

FE Ford main bearings are mostly interchangeable from year to year with a single notable exception. The outside diameter of the thrust bearing changed sometime around 1964. The later bearing will not fit the earlier block without modification. Many aftermarket bearings are now designed to fit both blocks.

My strong personal preference is to use either a 3/4-groove or 1/2-groove bearing on the mains. I actively try to avoid full-groove bearings. And I really dislike the idea of having an unnecessary "oil hole" in the lower main position, as is done on inexpensive bearing sets. I use the Federal-Mogul race bearings in all of my builds.

Bearing clearances are crucial to a successful build and should always be checked. Any of several methods may be used to verify bearing clearances. The best way to check these is with a dial bore gauge. A person working in a home shop might not have that tool available and will need to get by using Plastigauge. Any means of measurement is better than not checking. The clearance you are looking for is measured vertically since bearings change in clearance as they approach the parting lines. My preferred clearance range is between .0025 inch and .0030 inch.

Check Bearing Clearance–Micrometer

1 Use a micrometer to precisely measure each of the crankshaft's journal diameters.

2 Put the micrometer into a vise or holding fixture, with it locked to that measurement that you just made.

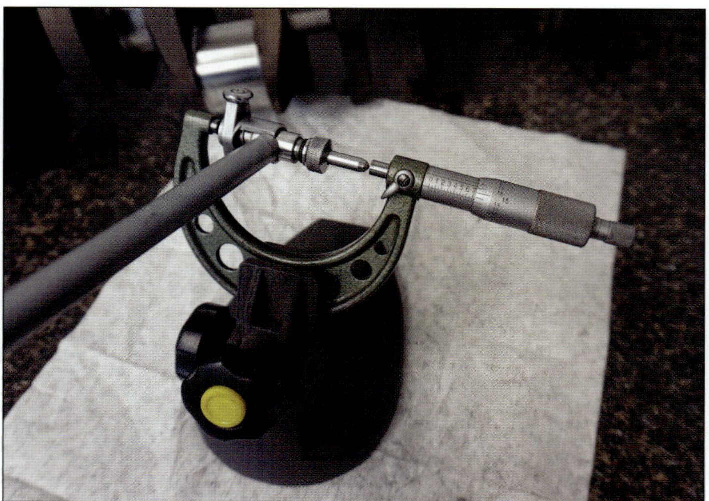

3 The bore gauge is then set to read "zero" when it is slipped into the micrometer.

4 You then take the "zero'd" bore gauge and slide it into the bearing. It will give you a precise measurement of the bearing's vertical clearance. Bearing clearances will vary around the diameter of the bearing, and are always measured at the vertical point.

Check Bearing Clearance with Plastigauge

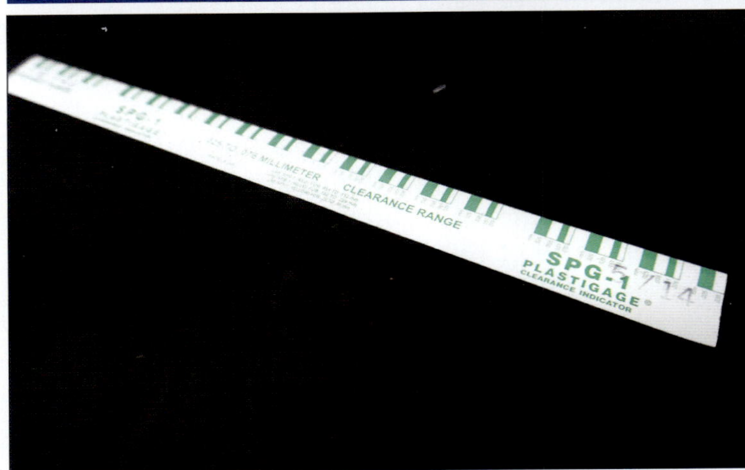

1 *If you do not have a dial bore gauge available, the next best thing is to use Plastigauge to verify clearances. This is a thin plastic strip that is compressed during a test assembly. You then read the thickness of the compressed plastic to get an indication of the assembly clearance. Not as good as a dial bore gauge, but far better than not checking.*

2 *To obtain a successful reading with Plastigauge you will set the lightly oiled crankshaft into the main saddles with bearing shells in place. Lay a single thread of the plastic on the journal to be measured, then install the main cap and torque it to specifications. It is important that you do not rotate the crankshaft at this point.*

3 *Carefully remove the main cap and you should have a clearly measurable strip of compressed flat plastic to make a reading. A thickness comparison chart is printed on the Plastigauge wrapper. Compare the width of the plastic strip to the chart to get a good approximation of bearing clearance. You will want to remove the plastic before continuing assembly. With some journal diameters you may have the option of alternate bearing sizes to adjust clearances. When available, it is perfectly okay to mix a +.001 or -.001 shell and a standard bearing shell on the same journal. In other cases you may need to have the crankshaft touch ground to gain necessary clearance if it is too small. If bearing clearances are acceptable, you may continue with assembly.*

Final Crankshaft Assembly

1 *You should always use a pre-lube or oil on the bearings where they meet the crankshaft during final assembly.*

2 *The back of the bearing that installs against the housing bore is left dry. No oil, adhesive, or sealer of any sort is desired in that location. But the crank surface always gets lubricated.*

3 *Install and torque bearing caps numbers 1, 2, and 4 first. The number-3 bearing on an FE engine is the thrust and requires extra attention. Lightly assemble that cap into place but do not fully tighten it. Use a large plastic dead blow hammer and give the crank a firm "knock" forward and backward to straighten and locate the bearing's thrust surfaces. Use a large screwdriver or pry bar to push the crank forward and hold it there while tightening the number-3 main to specification.*

4 *Now you can mount a magnetic dial indicator to the front of the block and set it up to measure thrust clearance. You should be able to smoothly push the crank back and forth in the block and see somewhere in between .006 and .012 fore and aft thrust clearances.*

5 *Cap number-5 includes the rear main seal. Rear seal problems are very common with FE engines. We do follow a particular procedure when installing the rear seal and cap to minimize problems, but even the best efforts are subject to some degree of possible leakage. The rear seal itself has a pair of semicircular rubber seals that are installed into the block and the main cap with the sharp edge always facing inward toward the block. The seal is always lubricated during assembly; it cannot be allowed to run dry against the crankshaft.*

6 Since the rear main seal area is the area that is least accessible after the engine is back in the vehicle, as well as most prone to leak, make sure this area is spotless before proceeding with assembly.

8 Slide the upper rear main seal half into the block, with the sharp edge of the seal facing in toward the crankcase. Put a small amount of oil on the surface of the seal that will be riding against the crankshaft.

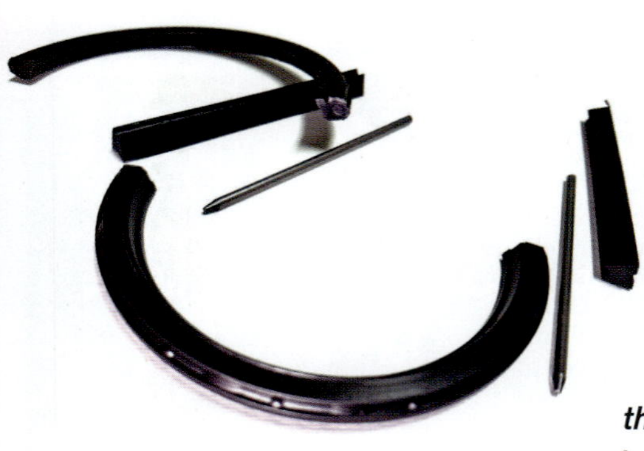

7 The Ford FE series of engine uses a conventional two-piece rear main seal. What is unusual is that the FE also uses a pair of vertical "side seals" that are slid into recesses in the main cap. I smear a very small amount of Motorcraft TA-31 silicone around the outer diameter of the main seal prior to installation.

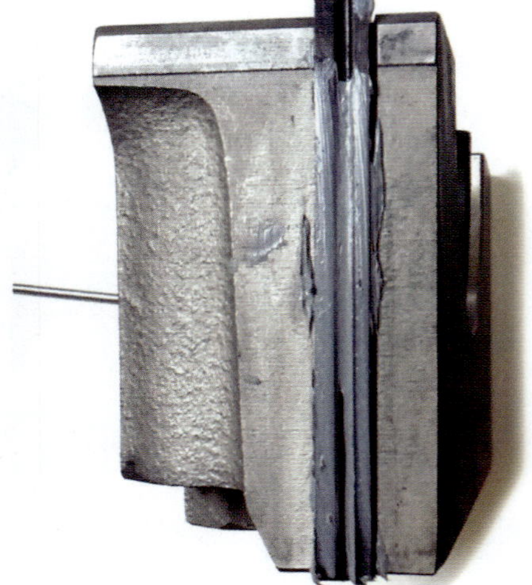

9 Instead of oil, use a small amount of Motorcraft TA-31 silicone as an assembly lubricant for both the rubber side seals and for the nails. The side seals are loaded into the grooves first, followed by a pair of nails that are driven into position in between the cap and the side seals to add pressure to the sealing surface.

10 Use a very thin film of the same silicone alongside the block sides and corners where the cap and side sealing surfaces meet. Slide the main cap into place as shown.

11 *Once the main cap is lined up, install the main cap bolts, and slide all the way into place, making sure the side seals remain flush with the top main cap surface.*

12 *Once the main cap is fully seated, insert the side seal pins and tap them into place carefully. Their primary function is to tighten the seal against the mating surfaces of the cap and block. Do not force the pins; if there is interference, pull the cap back off to see why.*

13 *The side seals and nails should be flush with or slightly below the oil pan gasket surface once everything is assembled. If done correctly you will see a tiny amount of silicone squeeze out between the main cap and the block along the sides and the cap register. This is the indication that you have completely sealed the voids at the rear of the block. Once the silicone has dried you can trim off any excess. The result should be a clean, flat surface for the oil pan gasket to seal against. A dab of silicone can fill a low spot and a razor blade can trim any protruding rubber.*

When completely assembled, you should be able to rotate the crankshaft smoothly by hand with only a bit of drag from the rear seal. If it takes a lot of force (or a wrench) to turn the crank, you have a bearing clearance or alignment issue that needs to be resolved before moving forward.

CONNECTING RODS, PISTONS, AND RINGS

Most factory connecting rods used in the FE engine family are a forged I-beam design with a floating pin. The most popular length will be 6.49 inches. Factory connecting rods are perfectly fine for most street-oriented engine projects. But we must remember that they are now 40 years old and most have seen considerable service time. At the minimum, the rods should be Magnafluxed for cracks, checked for straightness, and have the big end of the rod reconditioned. You should also replace both the fasteners and the bronze bushings. I will detail each of these processes and the assembly techniques in the following paragraphs.

Most common FE connecting rods are full floating and include a bronze bushing at the small end and 3/8-inch-diameter rod bolts for fasteners. The rods used in many 427 engines as well as in 428 Cobra Jet engines use a 13/32 fastener. If you choose to upgrade to ARP rod bolts, you will find that the threaded portion of both 3/8 and 13/32 ARP replacements is the same; this means that it's not worth chasing the CJ rod unless you are doing a restoration and really want those parts. Some

Shown is an original equipment piston and rod package from a 1969 Torino GT's 390 4-barrel engine. Of note is the basic rod and the cast piston used in even this moderately high-performance combination.

Factory connecting rods used in most FE engines are of forged I-beam design with 3/8-nut-and-bolt cap retention and a floating piston pin. Factory connecting rods are perfectly fine for most street-oriented engine projects.

On factory rods you may sometimes find odd-looking colors, including a copper dip. This is not a cause for concern, and these parts are perfectly usable for rebuild and continued service. The consensus is that Ford used these plating processes to bring parts right on the edge of tolerance into compliance with factory dimensional specifications.

Pictured is a Ford "LeMans" connecting rod with capscrew fasteners. Only installed in a few 427 and 428 Super CJ engines, these rods were also sold as over-the-counter performance parts. They are rare and not likely to be found in any normal FE.

The small, pin end of all FE connecting rods has a bronze bushing for the piston pin to ride in. These bushings are removed and replaced during connecting rod rework. New bushings are pressed into position, oil holes are drilled as required, and the bushing is finish honed to provide proper pin clearance.

high-performance engines used a capscrew "LeMans" rod, but these are rare and not likely to find use in a mild street-type application.

Connecting rod reconditioning, like crankshaft grinding, is a machine shop process; you take your parts in and let them do it. I will spend a short time describing the work they do but not dwell on it being it is not something you can do on your own in the garage.

The Magnaflux process will reveal cracks in the rod forging. It's not common to find these flaws in the factory FE parts, but still a very, very good idea to inspect before investing a lot of extra time and money in a set of rods.

Straightness can be verified by comparing and measuring the rod against a perfectly flat surface plate. The rod needs to show no signs of bending or twisting.

Once we have a verified good set of candidate forgings, we start out by resizing the big end of the rod. As connecting rods accumulate service time, they have a tendency to become oblong and out of round in reaction to the continuous stress they are under. Resizing them is a machine shop process that provides a perfectly round hole for the rod bearings to reside in. First

remove and discard the old bolts. Then use a cap grinder to remove a small amount of material from the parting surface of both the rod and cap, leaving the big end hole smaller than it originally was. The new rod bolts are pressed into place (they are a very snug fit), the rod cap is bolted to the rod and torqued to specifications, and the rod's big end diameter is honed on a dedicated machine to return it to its original and perfectly round diameter.

The small, pin end of FE connecting rods has a bronze bushing for the piston pin to ride in. These bushings are removed and replaced during connecting rod rework. New bushings are pressed into position, oil holes are drilled as required, and the bushing is finish honed to pro-

vide proper pin clearance.

With the reconditioning completed, the rods can be cleaned and set aside while we move on to the pistons and piston rings. They will be assembled together before installing them into our engine block.

Aftermarket Connecting Rods

As you inspect your existing parts, and price out the reconditioning, keep in mind the cost and availability of aftermarket replacement connecting rods. With brand-new

Aftermarket connecting rods, such as the Scat H beam products shown, are a viable alternative to reusing the original parts. While essentially mandatory for racing use, a new set of connecting rods can make financial sense when compared to the price of reconditioning older parts. The couple hundred dollars difference in overall cost is a modest portion of the overall engine budget.

parts available for only a couple hundred dollars more, sometimes it makes good financial sense to start over rather than rework old parts. Weigh the cost of repair, the nature of the build, and the impact on your budget when making the decision.

Pistons and Piston Rings

You have several choices to make when selecting the pistons for your engine, including size and dimensions, material, and a compression ratio. Each of these choices will have a significant impact on the performance potential and cost of the build.

Forged pistons are generally considered a significant performance upgrade, being much stronger and more durable under stress. They are also considerably more expensive. Most forged pistons allow the use of a more modern piston ring package, further enhancing performance.

Piston Dimensions

The first key choice is bore diameter as dictated by your engine block. Your machine shop will effectively make this decision for you when they determine how much they need to bore and/or hone the cylinders to get them prepared. Piston pin dimensions are decided by your connecting rod choice. OEM FE engines are .975-inch diameter.

Next will be the compression distance. This is a measurement from the centerline of the piston pin to the top of the piston. This selection is dictated by the deck height of the block, the stroke of the crankshaft, and the length of the connecting rod as measured from the center of the pin end to the center of the big end where the rod bearing resides. The formula is a simple one, as in the example below:

```
Deck height            10.170 inches
Rod length            - 6.490 inches
   Crankshaft stroke divided by two
(3.980/2)             - 1.990 inches

Compression distance (at zero deck
```
clearance) = 1.690 inches

In most cases we want to allow an extra .0050 to .0010 inch in clearance.

Material and Manufacturing Process

The material and manufacturing processes are interrelated, but separate, topics.

Piston choices for street builds are cast pistons or forged pistons. The cast pistons are what came originally in the vast majority of FE engines. Cast pistons are made from molten aluminum alloy that is poured into a mold. The castings harden as they are cooled; then they are machined into a finished part.

Cast pistons may be poured using either a standard or a hypereutectic alloy. The hypereutectic alloy is stronger and more durable, and can be considered an intermediate step, an upgrade from standard cast parts. Hypereutectic alloys have an increased amount of silicon in the aluminum mix, making them more resistant to wear in ring areas and skirts but also more likely to suffer

damage if your tune-up is way off and you get into detonation.

Forged pistons are generally considered to be a significant performance upgrade, being much stronger and more durable under stress. They are also considerably more expensive. Forging is a mechanical process that uses a large press and thousands of pounds of pressure to form a slug of aluminum alloy into a stronger, denser piston. I will generally move to a forged piston for all but the most budget-oriented projects, feeling that the added investment is justified. Most forged pistons will also allow the use of a more modern piston ring package, further enhancing performance.

Compression Ratio

Another key characteristic for piston selection is compression ratio. A number of online formulas are available to quickly calculate your compression ratio. This ratio is a comparison of the volume of air above the piston at the bottom of its travel to the volume at the top of its travel. The volume will be affected by

A modern forged piston will be lighter weight and use upgraded ring sizes and greater piston-to-valve clearance for use with performance camshafts. Forged pistons also offer a greater range of customization, allowing more options in every aspect of design and bore diameters.

A replacement cast piston is a low-cost alternative for the milder, street-oriented build project. These are proven very reliable and durable in lower-power and lighter-duty usage. The price is often far lower, sometimes by as much as 60 percent more than a comparable forged product.

Another advantage of a forged part lies in the potential for unique designs. This example is configured with a large-volume dish for reduced compression ratio and will be used with a supercharger.

the diameter of the cylinder bore, the length of the stroke, the volume of the cylinder head combustion chamber, and the shape and volume of the top of the piston. A piston with a dished top will thus provide a lower compression ratio, while one with a dome on top would be a higher compression unit.

Higher compression will provide more horsepower, but your available fuel will limit how far you can truly go. These days we are usually working with pump premium fuel with an octane rating between 92 and 93. This will limit the amount of compression that can be built into the engine before risking detonation and catastrophic failure. In general terms, a range between 9.5 and 10:1 compression is about as far as we normally want to go for a degree of safety in perhaps 80 percent of the street engines we build. In some applications you will want even less, in other applications you may choose to risk going higher.

Instead of fixating on a particular ratio number, consider the decision in terms of cylinder pressure. A higher load, lower RPM application that needs to idle nicely and have enough vacuum to accommodate power brakes will work best with a lower compression ratio. An application in a lightweight car with a big cam and manual brakes and transmission might be able to get away with a higher ratio since effective cylinder pressure will be reduced.

Piston Coatings

Various coatings have become very popular for pistons over the past several years. These fall into two general groupings: friction reducing coatings for skirts and protective coatings for domes and ring grooves. While no FE engine had a coating from the factory 40 years ago, they are very common in newer engines and have proven benefits.

Normally a dark gray color, the skirt coatings resist scuffing and are somewhat sacrificial. They will "wear

into" the skirt under high load or poor lubrication conditions, giving an extra layer of insurance against damage.

Although less common, the dome and ring groove coatings are most often found in extreme-duty applications such as those with superchargers or turbochargers. They are

A variety of specialized coatings are available in the aftermarket. These include skirt coatings for enhanced durability under marginal lubrication and resistance to scuffing. Even the original equipment manufacturers have adopted skirt coatings. They are a proven technology.

Piston ring selection is very important and your choice of material and face coatings from the many available options will affect both break-in and performance.

intended to reduce damage from ring micro-welding or light detonation.

Piston Ring Selection

Picking a set of piston rings involves a series of choices as well. These include base material, face coatings, tension, and dimensions.

Piston Ring Dimensions

Once again, most of these dimension choices are made for us by the block and by the piston manufacturer. The bore diameter is decided by the machine shop's work. The ring thickness is determined by the piston manufacturer since those parts must be compatible. Modern trends have been toward thinner and lighter

Top rings see the most heat and get the brunt of the cylinder pressure; probably not the best place to save a buck or two. Plain cast-iron rings should not be used in any but the most budget-oriented rebuilds. Premium rings will use either ductile iron or steel as a base material with moly coatings for far better strength, flexibility, durability, and resistance to detonation damage.

rings for reduced friction and better sealing, but you will need to go to a newer and more expensive piston to use that technology.

Piston Ring Materials and Face Coatings

Piston rings will be separated in the packaging and their position on the piston identified. Do your best to keep them in the correct envelope or carton. You really do not want to mix them up. Top rings with moly can often be identified by a darker center on the outside diameter surrounded by a brighter edge on the top and bottom of the cylinder contact surface. Second rings will have the same color overall. Most high-quality top rings will have a bevel or chamfer on the top inner diameter. Second rings

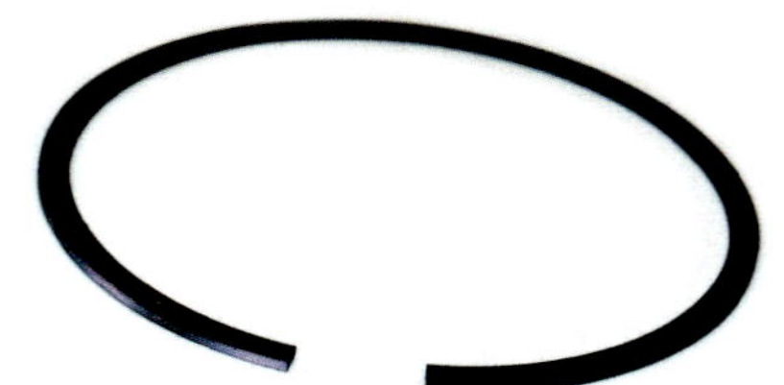

Second rings, not seeing the heat and compression loads of the top rings, are often made of cast-iron, which is adequate for the job. They serve as more of an oil scraper than as a top ring. Note that all upper and second rings will have a "this side up" marking denoting the proper orientation on the piston. This is especially critical on the second ring.

will often have a similar chamfer on the lower side of the inside diameter.

The top rings see the highest heat and loads. The least expensive rings are plain cast iron. These are not recommended as top rings in any but the cheapest of rebuilds. They will not last very long with unleaded fuels and high power levels. Piston rings are probably not the best place to save a few bucks.

Cast-iron top rings can be improved with a moly face coating. The moly works well to significantly improve durability and sealing. A basic moly ring should be considered the entry-level choice for any quality rebuild.

Premium rings will use either ductile iron or steel as a base material for far better strength, flexibility, and resistance to detonation damage. To adhere to the more flexible base material, the moly face on a premium ring is applied with a plasma process. This combination makes for a far superior, albeit more expensive, piston ring set.

Second rings are almost always plain cast iron. They see less heat and load than top rings do, and thus do not require the higher-strength materials or face coatings. Second rings are really oil scrapers, functioning as a squeegee to remove excess oil and keep a fine film of lubricant on the cylinder wall. They are not compression rings at all.

Oil rings occupy the third groove on the piston. They are most often a three-piece design. They will have a central expander, which is folded and perforated metal that serves as something of a spring. Above and below the expander are a pair of steel rails, which are pushed against the cylinder wall by the pressure of the expander.

Oil rings are most often an assembly of three different components. They include a pair of steel rails or rings that comprise the top and the bottom of the assembly, and their primary function is to scrape the oil that has splashed onto the cylinder walls. The center expander serves as a spring to keep the rails properly located.

Piston Ring Tension

Ring tension and the resultant drag accounts for a large percentage of the friction in an engine. Low-tension ring sets are usually focused on the oil ring and a reduction of ring cross-section dimensions. High-performance engines can significantly benefit from tension reductions, which come from changes in both dimensions and materials. Modern OEM engines run very-low-tension ring packages with excellent street performance. In most street-oriented FE builds, we still stick with the standard tension ring packages provided by the manufacturers. The ring package must coordinate with the piston design, and the modern and better ring designs are unfortunately not often compatible with the traditional replacement pistons we use in this book.

File Fitting

File fitting the end gaps on piston rings is a viable option for high-performance builds. Horsepower can be found in tailoring the gaps to the exact application, but a set of normal "drop-in" rings is per-

It is always a good policy to check piston ring end gaps before installing them into the piston grooves. In most cases, using the drop-in style of rings out of the box will be fine. To check end gaps, insert the ring into the cylinder bore, make sure it is square to the bore by pushing it into place with a ring square tool or the top of an inverted piston, and measure the gap with a feeler gauge. High-performance rings will use a file to fit sizing to optimize gaps for best performance.

fectly fine and functional in the vast majority of street projects. It is generally a good process to at least check the end gap by installing the rings into the cylinder bore one at a time, squaring them up with the appropriate tool or the inverted head of a flat-top piston. You then use a feeler gauge to measure the gap. A range between .016 and .030 is common.

File fit ring sets allow you to optimize the end gap. Ring manufacturers will each include a chart of desired gap specifications. A safe general rule on street engines is a minimum of .004 per inch of bore diameter on top rings. I personally prefer a larger gap on second rings, around .0055 per inch of bore diameter.

While you can file fit the gaps with a sharp, flat file, it's more common and much easier to use a dedi-

When installing the rings into the pistons, it is very important not to distort, twist, or bend the ring in the process. Using a dedicated ring expansion pliers, rather than twisting the rings in by hand for the top two rings, will help avoid damaging the rings during the installation process. The top two rings will have an identifier on them to indicate the upper surface. It is very important to orient them correctly for the best performance.

cated tool for the purpose. One tip is to concentrate on filing just one side of the gap, leaving the other side as a reference to help you keep the finished gap straight and parallel. It is critically important that you deburr and just "break" the edges of all surfaces you have filed. Ring groove clearance is extremely tight and the smallest burr or defect will cause the ring to bind up in the groove.

Installation and Fitting of Piston Rings

It is really a personal preference whether you wish to install the rings onto the pistons first or install the pistons onto the connecting rods first. Either approach works fine.

Piston rings should be installed on the pistons very carefully. You should use ring expansion pliers whenever possible for the top and second ring. Overexpansion of a piston ring can cause permanent deformation of the ring at the area opposite the gap. Winding a ring onto a piston will often leave a permanent spiral twist or bend in the ring, reducing sealing capability and potentially shortening its service life. If using a cast-iron ring, you risk putting a crack into the ring at a point opposite the gap, causing future failure.

Oil rings are installed by first winding the expander into the groove. Then the upper and lower rails are wound into place. On most expanders there will be a folded "foot" that holds the rails in position and pushes them out against the cylinder wall.

Both top and second rings will have some sort of identification on them to indicate the upper surface. This can be critical to getting good performance. The identifier can be a dimple, a circular dot, or the word "top" etched into the ring. Don't let that word confuse you; they use the word "top" on both upper and second rings.

Second rings come next. Often made from plain cast iron, these are the most fragile rings and the easiest to break. Expand them just enough to get them over the piston and place them into the proper groove.

Finally, install the upper ring in the same way, expanding just enough to get it on the piston. Some folks place a lot of value in particular gap locations during assembly, but I believe that as long as you do not line them up you will do just fine. I use a bit of light oil on the ring package and rotate the rings to be certain that they are free to move in the grooves.

Ring manufacturers include a chart of desired gap specifications. A safe general rule on street engines is a minimum of .004 per inch of bore diameter on top rings. While you can file fit the gaps with a sharp, flat file, it's more common and much easier to use a dedicated tool for the purpose. One tip is to concentrate on filing just one side of the gap, leaving the other side as a reference to help you keep the finished gap straight and parallel.

Twisting the rings into place is the preferred method for the bottom oil ring installation. Install the expander or center piece first, followed by the two rail pieces. The center piece acts as a spring to force the rails against the cylinder wall.

When properly installed, the rings rotate and flex smoothly in their respective positions. Make certain that the oil ring expander ends are not overlapped or bound up in the groove. Always use a small amount of oil on any metal part. Never install dry.

Installation of Piston Assemblies onto the Connecting Rods

We are going to assume that your piston manufacturer has properly sized the piston-to-pin clearances (this is actually a very good thing to have your machine shop inspect

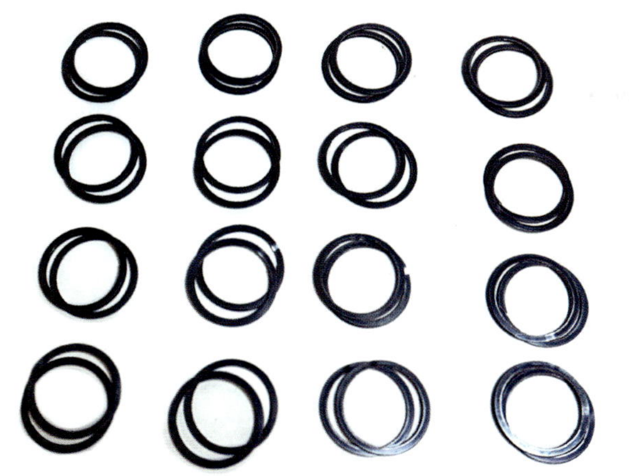

Ford FE engines all come equipped for free-floating piston pins, which are secured by Tru-Arc or Spirolox-type clips, as shown here. This means no press is required; you get to assemble everything by hand.

Connecting rods have a wide chamfer on one side of the big end as seen on the right, and a narrow chamfer on the other side (left). The wider chamfer faces the crankshaft counterweight, while the narrow chamfer faces the other rod on the journal.

and verify). Same for the equally critical piston pin–to–connecting rod bushing clearance, which should have been set when the rods were reconditioned.

Count your parts first. This can save a lot of worry later on if you find that there is a clip left over after putting everything together. Go one piston at a time; I start with number 1, the front piston on the passenger's side of the engine. Some pistons are stamped with an orientation arrow or notch to point toward the front of the engine. On Ford FE engine pistons with only two valve pockets, the pockets will always be positioned toward the center of the engine.

Connecting rods have a wide chamfer on one side of the big end, and a narrow chamfer on the other side. The wider chamfer faces the crankshaft counterweight, while the narrow chamfer faces the other rod on the journal. On the passenger's side of the engine the wide chamfer will face "forward." On the driver's side the wider chamfer will face toward the rear.

Ford FE engines all have free-floating piston pins. Pins are retained by Tru-Arc, round wire, or Spirolox-type clips. This means that the piston, ring, pin, and rods can all

The first step is to install one clip into each piston. This is done by twisting it in, somewhat similar to installing the oil rings. No special tools required.

Once you get the first edge of the clip in, simply maneuver the rest of the clip in a rotational fashion.

Here is what the first clip will look like after it is installed. Now is the time to properly orient your rods and pistons to make sure you assemble them properly before installing the final clip. Remember your rod chamfer orientation and piston valve relief location as you organize. Lightly lubricate the pin and pin bushing, and install the pin part of the way into the piston. Hold the small end of the rod next to the pin and slot it through. Verify your rod orientation and install the final clip as you did the first. Some pistons use two clips per side, so check!

To install the pistons into the bore, you need a ring compressor. Several styles are available, the most common being the compressing sleeve variety shown on the right, available for purchase or cheap rental at many parts stores. The tapered cone style (left) is easier to use, although you will need one to match your bore size, which is why they are not used as universally among home builders.

be preassembled by hand. No press is required. The process is simple, as long as you pay attention to a couple things. Tru-Arc rings install easily with the proper pliers. Spirolox take a little extra effort; they need to be stretched out a bit and then wound into position.

First install a single retaining clip in each piston. Lightly lubricate the pin, pin bushing, and pin bore. Hold the piston in your hand, start a pin into the pin bore, put a correctly oriented rod into position, and slide the pin through the rod's small end and into the opposite side of the piston until it stops. Double check your orientation (just to be certain) and install the remaining pin clip. Set each completed assembly on the bench and proceed to the next one in line.

Piston, Connecting Rod, and Ring Assembly Installation

At this point we are ready to complete the installation of the pistons and complete our short-block assembly. Again, start with the number-1 assembly at the front passenger's side of the block. Remove the rod nuts and separate the connecting rod bearing cap, making certain to maintain its proper orientation with the rod. Double check to make sure you have the piston and connecting rod properly oriented with the large chamfer

facing the crankshaft counterweight.

Give the cylinder a final wipe down with a clean rag and a light wipe of clean oil. Also put a bit of oil on the crankshaft's rod journal and massage a little bit on the piston skirts, the rod bearings, and the rings. You do not want to douse things in oil, but I do not assemble anything in an engine dry.

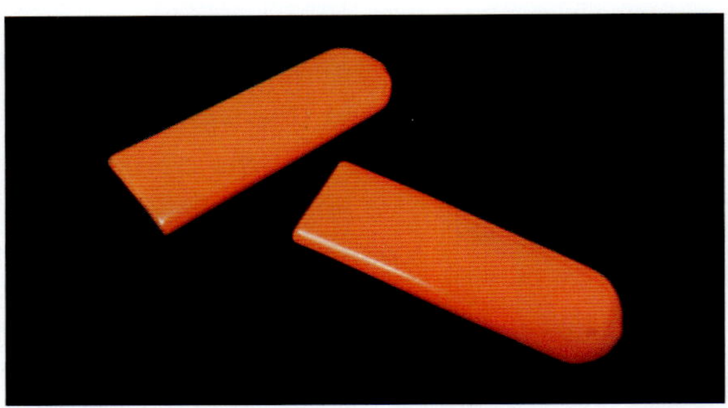

Before installing the piston and rod assembly into the engine bore, it is a good idea to put a protective covering over the threaded area of the rod bolts to prevent scratching of the crankshaft journals.

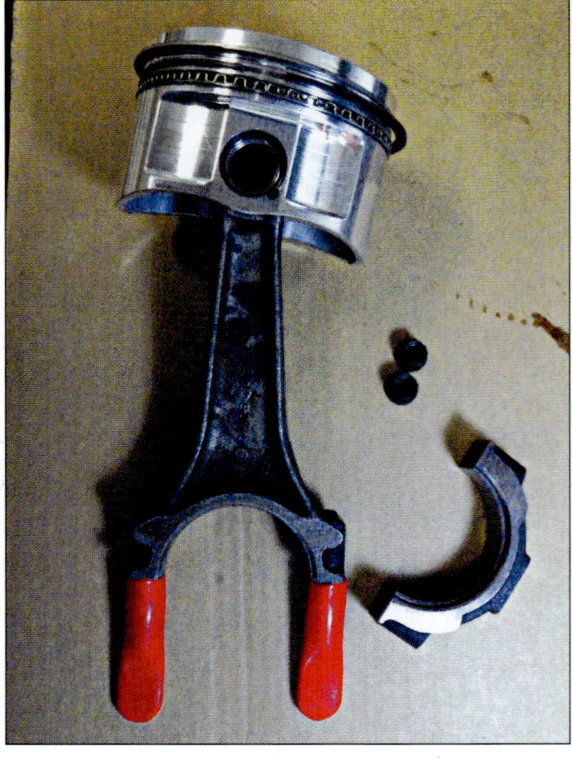

You can use dedicated protectors as shown here, or if none are handy, use rubber fuel hose cut to length instead. Either will do the job.

Massage a small amount of oil into the chosen cylinder bore for initial lubrication. Insert the piston and rod assembly into the proper bore, making sure the assembly is oriented properly.

Tap the piston down into the bore using a plastic mallet or the wooden-handle end of a hammer if in a pinch. Guide the rod through the cylinder bore, being careful not to let the rod touch or scratch the bore. Keep one hand above and one below for this procedure. Gently tap on the piston from above while guiding the end of the connecting rod around the crank journal. This should happen without much force.

Same procedure, different tool. You can tell with this ring compressor that you will wish you had a third hand when assembling, which is why the tapered cone–style compressors are so much easier to use.

Once the piston and rod assembly bottom out against the crankshaft rotate the engine to the oil pan side and install the proper rod cap. Double-check to make sure you have it properly oriented and slide the cap into position. Sometimes you'll need to use a plastic mallet to lightly tap the cap into its final position before snugging the rod bolts.

You will need a piston ring compressor to perform the installation. Several types are available, ranging from wind-up-style affairs, to flexible bands operated by a pair of specialized pliers, to simple tapered cones. Any of them will work. The tapered cones are perceived as more expensive since you need a dedicated tool for each size cylinder bore. I cannot overstate just how much easier it is to assemble the engine with the tapered cone though; they are well worth the investment.

If you are running connecting rods with traditional bolts it is a very good idea to use some sort of protective cover to keep the fasteners from accidentally damaging the crankshaft. These can be purpose-made

Torque the rod caps to the appropriate torque value. Be sure there is a light coat of oil on the fasteners for accurate torque readings. If you are using after-market rod bolts and caps such as ARP, be sure to use the torque specs supplied by the aftermarket manufacturer. You can use Plastigauge here to check rod bearing clearances just as you did with the main bearing previously. After confirming correct clearances, be sure to remember to clean and lubricate the bearing before final assembly.

boots or small pieces of fuel hose.

Install the chosen ring compressor onto the piston assembly, tighten it if necessary, and carefully lower the connecting rod down into the bore. You then guide the piston skirt into the bore until your ring compressor bottoms out against the cylinder deck. Make absolutely certain that the rings have not popped free of the compressor.

If everything looks okay, push the piston down into the bore using a plastic mallet or the wooden-handle end of a hammer if in a pinch. You want to guide the connecting rod around the crankshaft with your other hand as the piston descends. It might take a little force to get the assembly moving, but you should not need to beat things into place. If

it gets tight or stops sliding smoothly, *stop*. Sometimes a ring will pop free of the compressor. If that happens remove the assembly and start over. Keep hitting it and you will break a piston ring!

Once the piston and rod assembly bottom out against the crankshaft, you will move to the oil pan side and install that rod cap. Double check to make sure you have it properly oriented and slide the cap into position. Sometimes you'll need to use a plastic mallet to lightly tap the cap into its final position before snugging the rod bolts.

I normally install the opposite piston on that rod journal (number 5 in this example) before going to a final tightening of the fasteners. A quick note on fasteners and torque

values is appropriate here. You can ensure correct tightening on the highly stressed rod bolts in a number of ways: torque, bolt stretch, or "torque-angle" methods all allow you to arrive at a specific amount of fastener tightness. While the latter two are decidedly the most accurate methods, most FE engines will be assembled using the basic torque wrench. Done properly and using the values and lubricants specified by the fastener manufacturer, a simple torque wrench provides perfectly acceptable results. If you did a trial assembly and used Plastigauge to check clearances, remember to clean and lubricate the bearing before final assembly.

A good technique is to slip a feeler gauge between the two rods on the journal to keep things in alignment, and then to take each fastener to a snug degree of torque, roughly 50 percent of the specified final value, using the specified lubricant on the threads and underhead of the fastener. Then go back and tighten to the specified final torque for each fastener in a single smooth sweep of the wrench. Original equipment fasteners call for 45 ft-lbs of torque with engine oil, while an ARP replacement will be higher at 50 ft-lbs using their specific assembly lubricant.

Every time I install a piston assembly, I rotate the engine to get a feel for smoothness and friction. It should be relatively easy to rotate with a wrench with only one or two cylinders assembled. Each cylinder you assemble will add some resistance, but it should still go around smoothly. If it gets really stiff or difficult to rotate after installing a particular piston, you have something wrong and you need to find it and fix it now.

You absolutely must use some sort of lubricant on the crankshaft and bearings before torqueing everything into place. A special assembly lube can be nice, but even plain old engine oil will suffice.

An easy way to keep the rod caps in alignment during the torqueing procedure is to slip a feeler gauge between the caps as you are snugging them into place.

Rotate the engine after each piston is installed to be certain that it went into position properly. A correctly completed short-block will rotate smoothly with no binding or hard spots.

CAMSHAFTS, LIFTERS, AND TIMING

Cams are a topic of considerable discussion and debate coupled with startlingly little hard data to support opinions. Reasons for this are numerous, but among professional builders and racers a cam and engine package that really "works" is a competitive advantage that is closely guarded. The enthusiast that only does an engine every few years simply will not have the resources to make comparative tests where everything else is held equal.

FE Ford camshaft selection is similar to that for any other street engine. Your decision is based on exactly what you intend to do with the engine, what you are expecting out of it, and

what vehicle you are installing it in. If the basic selection is reasonably close, the differences between the appropriate cam choices for a given application will be modest. There are no true right or wrong cams within a rational range; they are going to be better in one characteristic and not as good in another. You need to decide which of those characteristics is most important to you and make a choice.

Camshaft selection is more about context within the build than about any one set of parameters. The *combination* of parts will define the outcome, not any single item. The characteristics of interest include idle quality, vacuum for

power brakes and accessories, the need for frequent inspection or service, the perceived drivability, and the desired power range.

FE Camshaft Design Specifics

The FE Ford engine does have a few design features that should be kept in mind. Most of these are well addressed by the various camshaft manufacturers and need only a cursory inspection when preparing to install your chosen piece. But they do need to be checked.

The FE uses a single dowel pin in the front of the cam as a locator and drive item. This pin has to be a snug, not press, fit; it should lightly bottom out in the drilled hole in the cam. The original Ford assembly had the pin press fit into the timing sprocket, but the aftermarket timing sets have long since abandoned that practice. If the preferred one-piece eccentric is used, the pin needs to be long enough to extend through the timing sprocket and come flush to the face of the fuel pump eccentric. On two-piece pump eccentrics, there

All FE camshafts (except for very early examples) share a common configuration, with five bearing journals and an integral distributor gear toward the front. Most will also have a grooved journal in the number-2 and -5 locations as required in solid lifter OEM engines, such as the 427 side oiler. Cams designed for flat tappet lifters will have a cast-iron core similar to the one shown here.

Check Cam Dowel Pin Length

If you work on more than one FE engine, you likely will run into different length cam dowel pins from time to time. The original pins from Ford were press fit; the aftermarket replacements no longer are. When using the preferred one-piece fuel pump eccentric, the pin needs to be long enough to extend through the timing sprocket and come flush to the face of the fuel pump eccentric. The less desirable two-piece eccentric requires a shorter pin since the inner part of the eccentric has a tang poking back toward the cam. ■

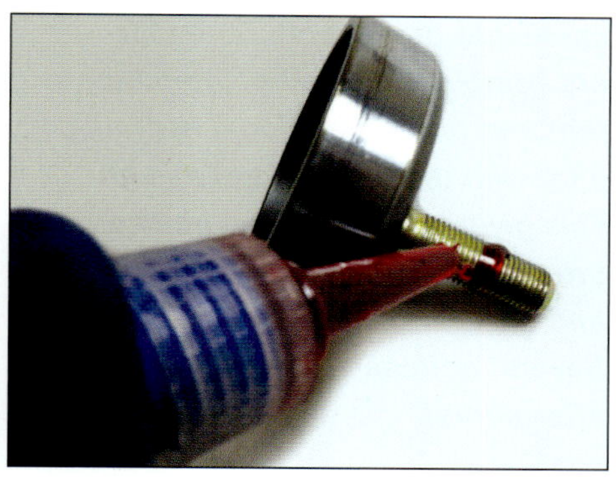

The front cam bolt hole on an FE is 7/16-14 thread from Ford. Most aftermarket cams will be the same, but we have found some with a 7/16-20 fine thread tapping instead that could cause damaged threads if unchecked. Be sure to use a quality thread locker along with an extra-thick washer when assembling the cam and fuel pump eccentric.

The front face of the cam runs against the thrust plate. The cam sprocket will be slipped onto the small snout that extends out from the front. Both of these surfaces need to be clean, flat, and free of any burrs or raised spots.

Camshaft Bearings

Cam bearings for FE engines are babbit lined, and most are manufactured by Dura-Bond. Standard FE blocks all use the same cam bearing set. The inside diameters are identical, but outside diameters get smaller as you go farther back into the block. This was done to permit automated installation in the factory. Side oiler blocks use a bearing set with different drillings to feed oil to the rocker arm assemblies but otherwise share the same dimensions.

Lifter Selection Criteria

First decide the lifter style to be used from the four basic options: hydraulic flat tappet, solid flat tappet (also referred to as "mechanical"), hydraulic roller, and solid roller. Each variation has advantages in certain types of use.

Keep in mind the fact that flat tappet lifters rotate against the cam by spinning in their respective bores; it is not a sliding contact. The bottom of a so-called flat lifter actually has a crowned profile that works in concert with a taper ground onto the cam lobe to promote the spin. You can see and check that radius by holding a pair of lifters against each other foot to foot in front of a light background. The profile will be readily visible. Lifters that are truly flat are worn out and should be replaced. When installing a flat tappet cam, it

is a small tab that extends rearward into the sprocket, requiring the pin be recessed from the sprocket's face enough to accommodate the tab.

We sometimes need to make a small spacer to slip in behind the pin when it's not long enough. Other times we end up shortening the pin. This needs to be measured and corrected before installing the pin into the cam because they can be a real challenge to remove without damage.

The front cam bolt hole on an FE is 7/16-14 thread from Ford. Most aftermarket cams will be the same, but we have found some with a 7/16-20 fine thread tapping instead. That could cause damaged threads if unchecked. And we often find that the tapped hole is rough and full of debris; it seems that the production

cleaning process misses that area regularly.

The FE camshaft has all five bearing journals at the same nominal 2.124-inch shaft diameter. The number-2 and number-4 cam journal will need a groove in their centers if you are running a 427 side oiler–type block. These grooves transfer oil for the valvetrain up to the cylinder head deck. The grooves are not required with the common blocks, such as a 390 or 428. Those blocks have annular grooves already machined into the cam bearing housing bores.

The distributor and oil pump drive gear is behind the front journal. We usually invest some time checking the teeth for burrs and rough edges. A few minutes spent with a wire wheel, small file, or sharpening stone will ensure longer gear life.

is *mandatory* that you use the proper cam break-in lube on all of the lobes and on the bottom of the lifters.

Roller lifters do not rotate in their bores; they have a bar linking two lifters together to prevent that. They use a roller that rides against the cam. The cam lobe is thus ground flat without a taper. The lifter's roller has a small radius ground into it to accommodate minor variances in lifter bore geometry.

Hydraulic Flat Tappet Lifters

These lifters are the original equipment in most FE engines. They are inexpensive, quiet, reliable, and perform perfectly well in most street-oriented applications where decent low- and mid-range performance is desired, combined with minimal maintenance. Sharing the same .874-inch diameter as other popular Ford lifters, the FE pieces are unique in that they do not have an

Hydraulic flat tappet lifters are original equipment in most FE engines and are the most common choice for replacement lifters as well. They are inexpensive, quiet, reliable, and perform perfectly well in most street-oriented applications where decent low- and mid-range performance is desired. It is vital that you use the proper break-in lube on all cam lobes and the bottoms of the lifters to prevent break-in failure.

oil hole in the pushrod seat. This is because the traditional FE does not oil through the pushrods.

Hydraulic lifters work by having an inner plunger that floats on a cushion of oil within the outer shell. Oil is fed under pressure into the cavity separating the two sections, and exits through the clearance between the inner and outer parts. A check valve prevents oil from exiting through the feed orifice. The controlled leakage allows the lifter to run with a small amount of preload that takes up any clearance in the valvetrain, and to compensate for wear. Most lifters will have roughly .100 inch of plunger travel between the upper retaining clip and the bottom of the cavity with the desired operating position centered in that range.

Hydraulic lifters for the FE fall into two groups: the stock replacement parts or the so-called race anti-pump-up styles. The only real difference between the lifters is the use of a heavier-duty retaining clip on the performance parts. This clip allows the performance parts to operate at or near zero preload without the risk of the lifter coming apart. While an improvement over standard parts from a safety perspective, they do not offer a dramatic performance advantage. Perhaps a couple hundred RPM can be gained from running at zero lash, but any wear or temperature-induced dimensional change will result in noisy operation and the need for adjustment. Avoiding those issues was the exact reason hydraulic lifters were installed in the first place.

Hydraulic Roller Lifters

These lifters combine the low maintenance features of the traditional hydraulic lifter with the

Hydraulic roller lifters are arguably the best option for a street engine but were never installed in any factory FE application. The engine was discontinued long before these came into production. Roller lifters combine the low-maintenance features of the traditional hydraulic lifter with the roller wheel design. This gives reduced wear, no need for break-in, reduced frictional losses, and access to enhanced cam profiles. Several options are available in the aftermarket.

roller wheel design. This gives reduced wear, no need for break-in, reduced frictional losses, and access to enhanced cam profiles. With the reduction in extreme pressure additives in modern oils (notably less zinc and phosphates) the break-in period on flat tappet cams has become a much larger issue than it was in the past. While successful break-in is certainly achievable without a lot of drama, the hydraulic roller retro-fit is a rather expensive way to completely bypass the issue.

Hydraulic roller lifters are arguably the best option for a street engine, but were never installed in any factory FE application. The engine was discontinued long before these came into production. The aftermarket has responded to the current demand by offering new parts that drop right into a normal

FE block. The lifters use a link bar design to prevent rotation, similar to that used in solid roller applications. All that is required are the lifters, the cam, and custom-length (shorter) pushrods. Since FE hydraulic rollers are newer parts, they can oil through the pushrods if desired. A true FE roller lifter will have the oil feed hole perpendicular to the roller axle. A lifter intended for a 460 engine will have the oil feed hole in-line with the axle and will potentially put too much oil up to the rockers.

The only downsides to hydraulic roller lifter installations are the effective limit on RPM and the initial cost. The lifters are large and heavy. Combined with the hydraulic design, they are best suited for moderate-RPM street engines with cams intended for a peak RPM at or below 6,200. In our dyno testing we've seen a pretty dramatic power drop-off once the valvetrain control is lost on hydraulic roller installations. Adding more spring pressure helps, but the hydraulic design can accommodate only so much pressure before creating new issues.

Solid Flat Tappet Lifters

This type of lifter is the best choice when you are trying to make horsepower on a budget. They will allow the higher 7,000-plus rpm levels that fall within the capability of the traditional FE block architecture. They are simple, inexpensive, and easy to install.

The solid flat tappets used in original FE engines have a dumbbell-shaped design, with a reduced diameter through the center of the lifter. There is no real advantage to these, but they're still considered to be the normal FE solid.

The downsides to a solid lifter

Solid flat tappet lifters are still a viable choice when building a high-RPM performance engine. They are simple in design, durable, and inexpensive as well. The only real downside, other than careful break-in, lies in the additional maintenance of periodic adjustments.

setup include the need for careful break-in and proper high-zinc oil and the requirement for periodic lash inspection and adjustment.

Solid Roller Lifters

These are the preferred choice for serious high-horsepower applications but fall outside the context of this book. The design allows virtually unlimited RPM potential, and they tolerate (and require) very high valve spring pressures. The cam profiles for

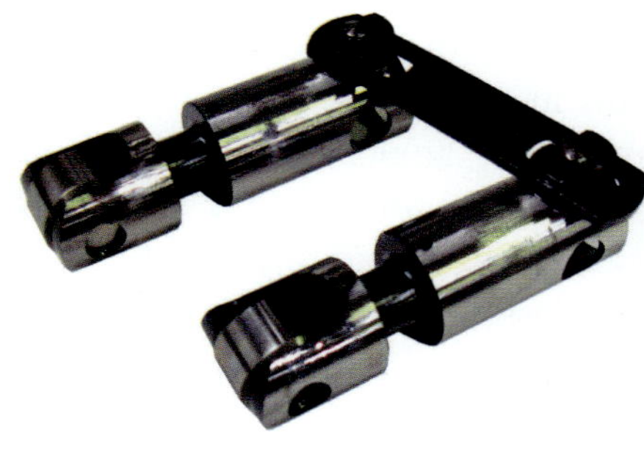

Solid roller lifters are the choice for extreme high-RPM performance builds using high spring pressures, aggressive cam profiles, and high lifts from radical cam lobe designs. The majority of solid roller options are oriented more toward racing applications than the typical street application.

solid rollers can be extremely aggressive with very high valve lifts; the roller can accommodate radical lobe designs.

Cam Specs: Lift, Duration, and Lobe Separation Angle

Cam specifications are given in valve movement increments. Most camshafts are selected based on the lift and duration numbers provided by the various camshaft manufacturers. They provide gross valve lift, an

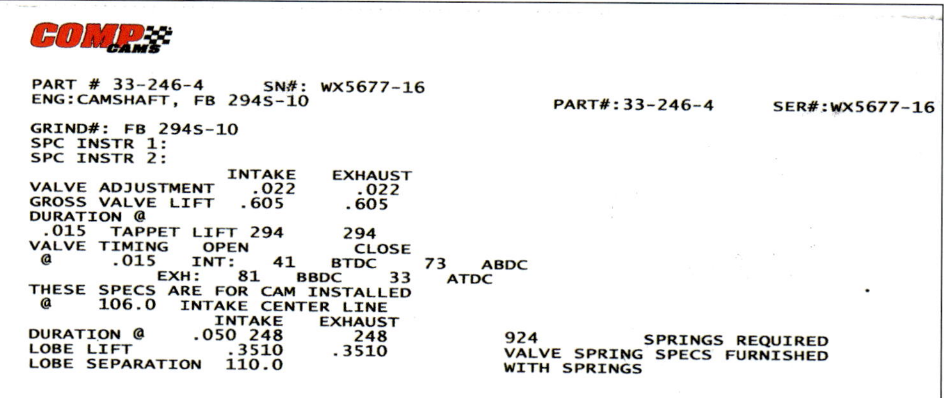

```
COMP CAMS

PART # 33-246-4        SN#: WX5677-16
ENG:CAMSHAFT, FB 294S-10                          PART#:33-246-4        SER#:WX5677-16

GRIND#: FB 294S-10
SPC INSTR 1:
SPC INSTR 2:
                     INTAKE      EXHAUST
VALVE ADJUSTMENT      .022         .022
GROSS VALVE LIFT      .605         .605
DURATION @
 .015   TAPPET LIFT 294          294
VALVE TIMING      OPEN            CLOSE
  @       .015    INT:    41    BTDC    73    ABDC
          EXH:    81    BBDC    33    ATDC
THESE SPECS ARE FOR CAM INSTALLED
  @     106.0   INTAKE CENTER LINE
                   INTAKE      EXHAUST
DURATION @       .050 248        248              924             SPRINGS REQUIRED
LOBE LIFT             .3510      .3510            VALVE SPRING SPECS FURNISHED
LOBE SEPARATION  110.0                           WITH SPRINGS
```

You will receive a cam card with any quality performance camshaft. It will provide a wealth of information, including duration, lift, lobe separation angle, and the desired installation position relative to the crankshaft.

advertised duration number, durations at .050-inch lift, and the lobe separation angle. Good and useful information, but only a small part of camshaft lobe design. Professional cam designers work with lobe acceleration rates and opening/closing events to arrive at the desired valve motion. The duration data and lobe separation points are thus an output from the effort, rather than an input. This means that, while directional information can be derived from those specifications, you cannot assume that they are the entire story. Context is the key in comparing similar cams.

Lift is the maximum distance that the valve is moved from its seat, measured in thousandths of an inch. Gross valve lift in increments of an inch is derived by multiplying the cam lobe lift times the rocker ratio. This definition includes a lot of assumptions. First is that rocker ratio is accurate, which is rarely the case. Next is that it does not reflect the impact of valve lash. Last is that the valvetrain will always have some flex or deflection. It is not unusual to see considerably less actual lift when it is measured at the valve. Checking lift at each spot in the valvetrain will tell you where the loss occurs by comparing measurements to the mathematical ideals. Reducing deflection through use of stronger rockers, heavier-duty pushrods, and better mountings will yield a significant improvement in both performance and durability.

Having the valve lift beyond the best flow range of your heads is okay in a race engine, providing the port flow does not get turbulent or back up at the higher lift. This is because of the next variable, time, as in the amount of time that the valve can be held open at that peak level for maximum flow through the port.

Duration is the amount of time that the valve is open, as measured in crankshaft degrees. Since the crankshaft turns twice as fast as the camshaft, the number of degrees quoted will be comparatively large. We have 720 crankshaft degrees per one full turn of the cam. Advertised duration numbers are of nominal value since there is no real standard for the measurement; some cam companies use .002-inch lift for this measurement, others use .006 inch, others a zero base. The "duration at .050 lift" specification came about as a method of industry standardization, allowing comparison of cams from different suppliers. In the crudest of simplifications, the duration determines the operating RPM range of a camshaft in a given application.

Duration numbers, both advertised and at .050-inch lift, are useful for guidance but not definitive when comparing cam profiles. Various software programs can get you "close" on basic cam selection using these values. But the only real way to compare cam lobes is by plotting the full range of valve motion throughout the lift curve. You can easily find two cam lobes with comparable lift, advertised and "at .050" numbers, that will be significantly different at every other point in the range of motion. Comparing cams would be better done with a graphed curve showing duration at, for example, .100 inch, .200 inch, .300 inch, .400 inch, etc. None of the large cam companies currently provide that level of detail at the consumer level.

Lobe separation angle refers to the difference between the intake and exhaust lobe centerlines as expressed in degrees. This specification is properly considered an "output" of the chosen valve events rather than a design "input." That noted, common assumptions are that cams having a wide angle of 112 or more degrees are street oriented for a broad power band and smoother idle, and that separation angles of 106–108 are race car material with stronger peaks and a narrower power band. These ideas contain a grain of truth since a cam designer may use the same lobes and just move them around to change the tuning behavior, but they may also use completely different lobes to get the performance characteristics they desire, changing the angle values dramatically without changing the tuning behavior.

Intake Lobe Centerline is expressed as a number of degrees before crankshaft top dead center. This specification is used to install the cam at the proper position relative to the crankshaft, as desired by the cam designer. Cam lobes often are not symmetrical, so the centerline is not going to be at the point of maximum lift. Instead, you find the centerline by locating a point .050 inch lower than peak lift on both the approach and descending sides of the lobe and splitting the difference. Installing the cam at the centerline position specified by the manufacturer should result in the opening and closing events occurring per specification. When cross-checking the opening and closing events, you must use the proper lifter (a solid lifter will give bad data if used for checking a roller cam); the lobe's flanks are completely different even though the event specs may appear to be similar.

Advancing or retarding the cam will change the opening and closing events relative to the crankshaft and alter the installed intake lobe centerline. It will not change the lobe separation angle, lift, or duration values, as those are ground into the cam itself. Altering the installed position by a few degrees can have either a modest or a significant impact on the way the engine runs. This is a tuning change with results that cannot be easily predicted. You have to try it.

So How Do I Pick a Cam?

The cam selection process can take a few different paths: You can open a catalog and find the one that most closely matches your desires based on the manufacturer's information. You can call the cam company's tech lines and get more detailed recommendations from the guys that answer such calls constantly. You can consult with your friends, engine builder, or machinist to benefit from their past experiences, or you can recruit a professional cam designer to create a cam profile matched to your specific needs.

None of these are wrong, but some methods are going to increase your chance of success. The engine builder, cam designer, and tech department are all likely to have considerable experience with the kind of questions that will arise. The designer will work from your combination to generate a profile, while the tech group will try to match a pre-existing catalog profile to your application. The cam designer will likely end up with the stronger combination, but that increment of power comes with a cost that is potentially too high for a more modest street build.

We'll provide general guidance based on what has worked for us in various FE engine builds. As noted earlier, this is not to be taken as definitive but more as directional. Your results

In comparison to the flat tappet cam shown earlier, this is a roller camshaft. Most roller camshafts will be made from steel. The machined finish visible between the lobes readily identifies them. Steel cams require the use of a compatible distributor gear.

will certainly vary depending on the engine characteristics, including displacement and compression, along with car variables such as transmission, rear-end gearing, and your own perceptions and expectations regarding drivability, sound, and power.

Working with the common catalog cam profiles, we can make some very basic recommendations. These are dramatic simplifications, but provide a starting point. On a 390 engine keep the duration at .050 down to a range between 215 and 225 degrees for a mild idle, and 230 to 240 degrees for a choppy one. As you go up in displacement, you can go up a similar amount in duration and retain comparable driving and idle characteristics. A 10-percent or greater increase in displacement to a 445 can easily handle a cam with 235 degrees of duration at .050 lift with very good street manners. It's something of a nonlinear sliding scale though, and even a 482-ci engine will want only around 250 degrees for a comparable performance range.

For lift we try to keep the milder street combinations at or below the .600-inch lift range. The reasons for this are twofold. First is that the Edelbrock heads we use are delivered

with a spring and retainer package oriented around a .600-inch peak lift. Second is that keeping the lift within this window reduces likelihood for geometry issues or pushrod-to-intake interference.

One statement that holds true is that a larger-displacement engine can use a larger-duration cam with good results. Since most FE cam catalog recommendations are based on standard displacements, the now popular stroker engines can comfortably go a step up in cam sizing without ill effect.

On 390-based 445-ci stroker engines we can use the old-school Comp Cams 282S and 294S solid flat tappet cams as examples. With 236 degrees of duration at .050-inch lift, the 282S delivers a noticeable but not really choppy idle at around 750 rpm. Gross valve lift is noted as .571; subtract .022 lash to get a net lift of .549 inch. This is low enough to use the springs included with the popular Edelbrock heads, although we usually change out for a double spring. Power peaks at about 475 at 5,500 rpm with 503 ft-lbs of torque.

Going to the Comp Cams 294S grind, with 248 degrees of duration at .050-inch lift and .605-inch gross lift will get more power and a higher RPM peak of around 6,000. With +/- 20 more horsepower and more torque throughout the power range this seems like a winning package, and it is. But it comes at the cost of a much choppier 850–900 rpm idle, reduced low-RPM response, and reduced idle vacuum that might not be happy with power brakes.

Now go to a 482-inch 427-based stroker and the situation changes dramatically. That same 236 at .050 282S cam we started with now delivers 537 hp at 5,700 rpm. The torque also moves up to 574 at 4,200 rpm. Idle is nearly smooth at 700 rpm. You get excellent driving characteristics in a much tamer-sounding package. Until you hit the throttle and the torque takes over. But it runs out of steam pretty early for such a racy short-block, making this a great combination for a Galaxie cruiser or a pickup truck.

Going back to our original discussion on selection, there are no perfect or right answers. Engines cannot read. Use the best knowledge and the most experience you can acquire and afford and then make an educated camshaft choice. Then you're going to have to go and try it in your car. Only then will you know whether it works to your satisfaction.

Cam Thrust Plate and End Play

The camshaft in an FE is retained by a cast-iron thrust plate that is surface ground on both sides. This plate is mounted to the block with a pair of .750 fasteners. On original engines they are Phillips-head or button-headed hex drive–type screws. A normal hex headed fastener will usually clear the aftermarket double roller timing sets, but you should check to be certain.

The thrust plate is sandwiched between the cam sprocket and the camshaft, allowing about .005 inch in cam end play. This should be measured and verified. The plate can be polished to gain some, or the

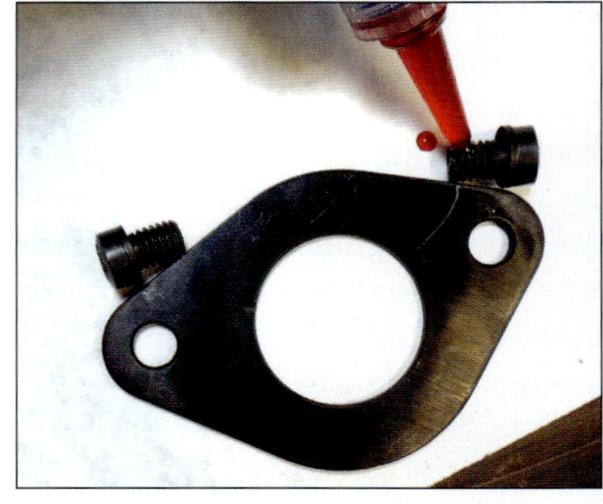

Be sure to use thread locker on the fasteners when installing the cam thrust plate. We prefer the red Loctite in this location, and both surfaces must be very clean for it to work properly.

sprocket altered to reduce it. Original thrust plates can often be reused, but they do get damaged and worn from debris. Replacement thrust plates are available.

Timing Sets

If your collection of used parts includes a C-shaped spacer washer that originally went between the

This is your typical cam thrust plate mounted in place; it is surface ground and fastened by two .750 fasteners. It is sandwiched between the cam sprocket and the camshaft. The original attaching fasteners have a large number-4 Phillips-head screw design. These are frequently replaced with a hex head fastener in service.

The cam end play should be around .005 inch, which should be measured and verified. If replacing the factory fasteners with aftermarket ones, be sure to double-check for interference with the timing sprocket.

cam sprocket and the camshaft, discard it. All the replacement timing sets have that spacer integral to the sprocket and the spacer washer is no longer required.

FE fuel pump eccentrics can be either a one- or two-piece design from the factory. The replacement two-piece eccentrics are poorly designed and should be avoided. You can pur-

For reliability and durability, stay away from the aftermarket two-piece fuel pump eccentrics. You can use this piece from Ford Racing designed for the 385 series engines, but be sure to check timing cover clearance, as you may need to trim to fit.

chase new one-piece eccentrics from Ford Racing; they are designed for a 460 with a 7/16 bolt. They are a little too thick for an FE and can sometimes hit the timing cover, a problem easily addressed by trimming them down in a lathe.

The bolt that attaches the cam sprocket needs to be a high-quality Grade 8 or better fastener. We use an ARP part. The washer behind it needs to be extra thick and strong to withstand the torque. A hardware store washer will deform, causing the bolt to come loose. The washer also needs

to be large enough in diameter to cover, at least partially, the locating dowel, preventing it from coming out.

The crankshaft sprocket on an FE is a very light press fit and should tap into place with minimal force using a piece of tubing and a plastic mallet. Factory FE crankshafts have a long key that inserts into the crankshaft as a locator for the sprocket before installation.

Timing sets are available in numerous configurations and many levels of quality. The most basic timing sets are a traditional link style single row, which is very similar to the original parts. Replacement cam sprockets have abandoned the original aluminum hub and nylon tooth material for a stronger and more reliable cast-iron material. They all have a dot for indexing the

Timing chain and gear sets come in multiple styles. Often from the factory the cam gear came in aluminum with nylon gear teeth, in a quest to make the assembly quieter. These were problematic because the nylon became brittle prematurely, broke off, and fell into the oil pan. This often caused the chain to jump because of the newfound clearances, and the teeth often clogged up the oil pump pickup screen, restricting oil flow. You not only had to replace the chain and gears, you also had to drop the oil pan to get the nylon teeth out. Modern single-row sets are made of cast-iron and are much less problematic.

A popular upgrade is a double roller chain, which uses two rows of teeth on the sprockets and a different type of chain. Roller chains are generally considered stronger and more durable in performance applications. Double-roller sets are available with multiple crankshaft key slots that allow degreeing the camshaft to optimize cam timing. Crankshaft sprockets are always steel parts and will have a dot to indicate the top dead center position.

Installation of the timing set is fairly straightforward. Rotate the number-1 piston to top dead center (TDC) using a piston stop or dial bore gauge.

Slide the crank sprocket most of the way into position but don't push it all the way back yet. Use the cam sprocket and dowel as a handle to rotate the cam into the position where the dots of the sprockets line up with each other. On a basic combination you are nearly done. Drape the chain around the cam sprocket and wrap it around the crank sprocket as you put the cam sprocket into position. Once you get the dowel started and the center of the cam going into the sprocket, you will alternately tap and push both sprockets into position. You want to minimize any stress on the chain, so go slowly in small increments.

Once everything is seated, install the fuel pump eccentric, the cam washer, and the cam bolt. If you are using a basic dot-to-dot approach, you can finish up by putting some red Loctite on the cam bolt and tightening it to specifications. If you are intending to degree the cam, wait on the thread locking compound until you're certain the cam timing is as desired.

number-1 location for the cam. The popular upgrade is a double roller chain, which uses two rows of teeth on the sprockets and a different type of chain. Roller chains are generally considered stronger and more durable in performance applications. Double-roller sets are available with multiple crankshaft key slots, which allow degreeing the camshaft to optimize cam timing. Crankshaft sprockets are always steel parts, and will have a dot to indicate the top dead center position.

I admit to a personal preference for Cloyes timing products and lean toward higher-end double roller sets for my own use. They fit properly, they have optional sets (though rarely necessary) for reduced center-to-center distance, and I like the flexibility of the multiple keyway feature.

Installation on a basic set is pretty straightforward. Rotate the number-1 piston up to top dead center using a piston stop or a dial indicator. The stop is more accurate, but on a street engine the gauge will get you plenty close enough. I often mark that point on the back of the crankshaft and the block with a Sharpie; it comes in handy while checking things during assembly. Slide the crank sprocket most of the way into position but don't push it all the way back yet. Use the cam sprocket and dowel as a handle to rotate the cam into the position where the dots of the sprockets line up with each other. On a basic combination you are nearly done. Drape the chain around the cam sprocket and wrap it around the crank sprocket as you put the cam sprocket into position. Once you get the dowel started and the center of the cam going into the sprocket, you

Degreeing the Cam

On a race engine we use cam timing to optimize performance, and may change it to tune the combination. On a street build it's more of a checking process, making sure that the cam and timing components are as intended. To check cam timing you need a dial indicator with magnetic mounting, a degree wheel, and the timing card that came with your camshaft.

Once satisfied with the cam timing, you can remove the degree wheel, apply thread-locking compound to the cam bolt, and torque it to specifications. I usually use an oil can to squirt some lube on the sprockets, the chain, and the thrust plate areas at this point. ■

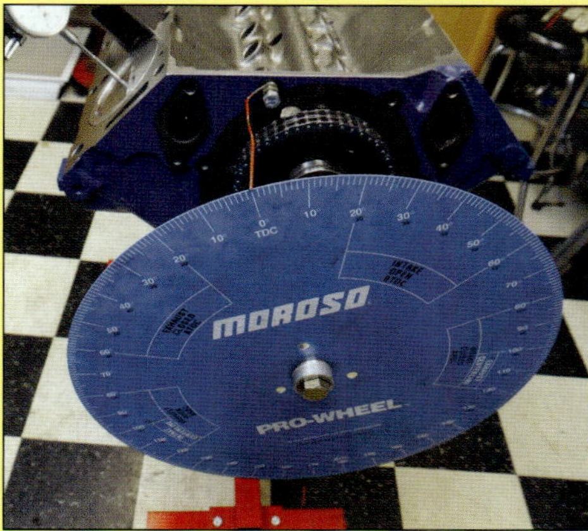

With the number-1 cylinder at top center you want to mount the degree wheel to the snout of the crankshaft. Fabricate a pointer that indicates the top center location on the wheel (a piece of welding rod or coat hanger gets the job done).

Drop a lifter into the number-1 intake position. That is the second lifter back on the passenger's side of the engine. A solid lifter is best so that it will not compress as the gauge moves against it. Set up the dial indicator so that it reads directly as the lifter travel upward. We are intending to locate the center of the cam lobe. Since many lobes are not symmetrical, we want to locate the point on the degree wheel that is .050 inch before the highest lift, and the point that is .050 inch after that highest lift. We will split the difference and thus find the intake lobe centerline. It will probably take a few rotations of the engine to get a clear and repeatable reading if you have not done this before.

Once you have found the current installed position, compare it to the specifications on the cam card. In most cases it will be close. A multiple keyway timing set gives you the option of advancing or retarding the cam to either better meet the cam card specifications or to alter the combination a bit. In very simple terms, advancing the cam will accentuate lower-RPM performance at the sacrifice of some top end power. Retarding the cam will soften the low end and possibly gain some peak power. While useful, cam timing changes will not make up for a significantly incorrect cam selection. A 4-degree change will be apparent, while a 1- or 2-degree change might be very subtle.

Slide the front oil slinger onto the crankshaft. This is a formed sheet-metal disc that directs splashed oil away from the front seal. Although a "correct" and good thing to use if you have one, we have built many engines without one and have seen no noticeable problem. If you have it, run it. If not, I would not be terribly concerned.

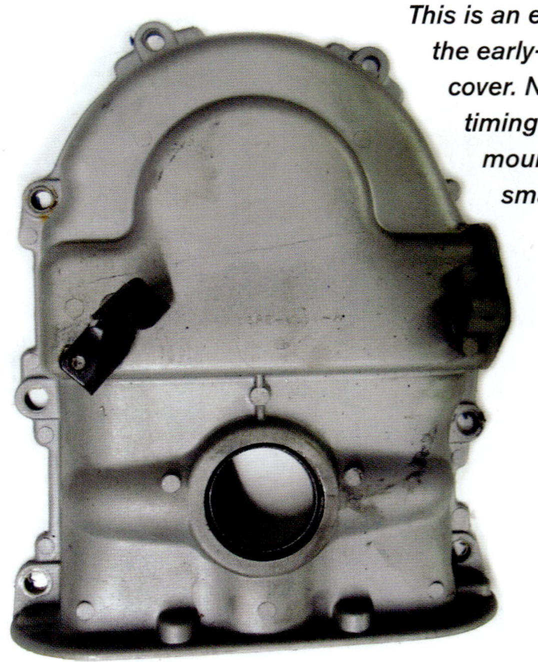

This is an example of the early-style timing cover. Note that the timing pointer is mounted with two small fasteners.

This would be considered a late-style timing cover. You can see the extended fastener hole on the left, which allows a longer, 3/8 bolt to mount the timing pointer at that end.

This example will accommodate both early and late pointers. Note the extra holes for both small fasteners as well.

will alternately tap and push both sprockets into position. You want to minimize any stress on the chain, so go slowly in small increments. Once everything is seated, install the fuel pump eccentric, the cam washer, and the cam bolt.

If you are using a basic dot-to-dot approach, you can finish up by putting some red Loctite on the cam bolt and tightening it to specifications. If you are intending to degree the cam, wait on the thread-locking compound until you're certain the cam timing is as desired.

Timing Covers

Two popular timing covers are used in most Ford FE engines, and the few others are rather uncommon. Really old engines used a sheet-metal cover. We won't spend much time discussing those, other than to note that they exist. Heavy-duty truck and some marine applications

Important!

Carefully install the crank seal, making sure the lip of the seal points toward the rear of the engine. Make sure you don't lose the spring in the process; it is important in maintain the proper pressure on the rubber seal against the crankshaft hub.

If you have the equipment, use a press to install the front seal. Most home builders do not have this luxury, so using a large socket as a drift and tapping it in lightly with a hammer is the next best method.

After the seal is installed, double-check and make sure the spring is still intact. Now you are ready to install the cover to the engine block gasket.

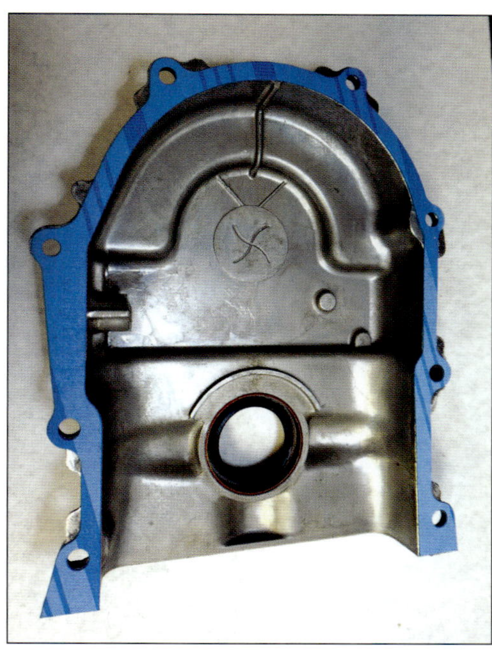

Use a very small amount of gasket sealer to hold the gasket in place, and make sure the gasket does not shift or move while you mount the timing cover on to the engine block.

Mount the timing cover on to the block and only loosely install the mounting bolts at this time. Now install the damper spacer. The spacer should slide smoothly into place, using a bit of anti-seize compound on the inside diameter.

used a cast-iron cover. Again, just be aware that they are out there. The restoration- or marine-oriented folks building applications requiring either one will know what they are, why they need them, and how to install them.

Most of us are building passenger car and light truck engines. The original timing covers for these are die-cast aluminum. The restoration folks keep track of various part numbers for them. The earlier variation used two small self-tapping fasteners to mount the timing pointer. Later versions (if you can consider 1968 as "late") use a longer 3/8-inch fastener at one side and retain a single small fastener for timing pointer mounting. As long as you use a matching pointer, the two covers are functionally interchangeable. The lower front

With the damper spacer in place, it acts as a centering device for the timing cover and front seal to ensure proper sealing on the crankshaft. Now it is safe to tighten the timing cover bolts to specification.

The damper shown here was used on a variety of FE engines. The pulley groove is integral to the outer ring of the damper. If you want or need to reuse one of these, be certain that it is in excellent condition. Any deterioration of the rubber will be a sure indication of impending failure. Driving the pulley with the damper might have seemed like a good idea at one time, but it is proven to be a bad design and something to be avoided in any performance application.

Shown is a Power-Bond replacement damper. Most after-market dampers are fully interchangeable among the various FE applications. All passenger car FE dampers are internally balanced, so there are no issues with replacement from that perspective. Most of them will accommodate the variety of factory pulley setups as well. With the original dampers being 40 years old or more, a replacement is a great idea for all but the most budget-oriented build.

of the timing cover bolts onto the oil pan and forms the seal there.

The only areas of concern to check on timing covers, and repair as needed, are the pointer mounting threads, the 3/8-16 threads on the two fuel pump mounting locations, and the 5/16-18 threads on the oil pan surface. All tend to see a lot of damage from service.

Install a new front seal and mount the timing cover. I use a very small amount of gasket sealer to hold the paper gasket in place during assembly. Use a bit of Teflon sealing paste on the timing cover bolts that go into the water jacket and into the lifter valley. Run the fasteners into place but leave them a bit loose until the next step.

Next install the damper spacer. This is a cylindrical sleeve that provides the inside surface for the front seal to ride against. It needs to be very smooth with no visible wear or it will leak. Repair sleeves are available, and new damper spacers are fairly inexpensive. On passenger car and light truck applications the damper sleeves are all identical with the single notable exception of the

428 Super Cobra Jet, which used an additional counterweighted spacer. The spacer should smoothly slide into place; I use a bit of anti-seize lubricant on the inside diameter, and it will index onto the same keyway that located the lower timing sprocket.

Use the damper spacer to center the timing cover around the front seal, and snug up the fasteners. Trim any gasket material that extends below the pan rail surface.

Dampers

FE dampers (also referred to as harmonic balancers) are all internally balanced. There are a couple fairly popular factory variations for passenger car use, and plenty of date code iterations for restoration guys.

Some earlier versions used a combination damper and lower pulley that ran the belt on the outside diameter of the damper. It probably seemed like a good idea at the time, but unless you are working in a restoration environment, I would avoid these. Attrition of the rubber in a damper is bad enough with-

out the added loads of belt-driven accessories.

Most factory and all aftermarket dampers will be more conventional, with all belt drive pulleys fastened on the front of the mounting hub. Outside of the rare high-performance items and the aforementioned combination pulley/damper parts, the FE Ford dampers will have timing marks engraved onto the damper outside diameter. If the rubber in a damper has degraded at all, replacement is highly recommended.

A wide array of suppliers provide aftermarket replacements. Most suppliers will cover all the popular FE applications with a single part number. On stock or near-stock engines I am partial to the PowerBond product, while high-performance builds will get a Romac or ATI part. You should always check and verify the TDC markings on your chosen damper against the timing pointer, as variances do occur. The ATI in particular is a racing-specific part and has timing marks in a unique location that need to be accommodated.

The damper presses onto the crankshaft snout. Some suppliers

With the cam installed and degreed, the timing cover mounted, and the dampener in place we now have a completed short-block awaiting heads and valvetrain!

have a specified interference dimension and expect you to hone the damper to get that fit. It definitely should not just slide on, and should not require excessive force either. Assuming you are in good shape dimensionally, you will first install the required 1/4-inch key stock into the crank snout. It's just a short piece and should not extend beyond the end of the crankshaft. Use anti-seize lubricant on the crankshaft's snout. Start the damper onto the crank after getting it lined up with the keyway. A few moderate taps to the center hub with a plastic face dead blow mallet will get it started on far enough to remain on the crankshaft

as you continue. Do not beat on the outside inertia ring of the damper, and do not use a metal sledge to force things on.

Once started, you can use a dedicated damper installation tool to finish pushing it into place. Or you can use the damper bolt to pull it in. There is a real benefit to using the proper tool, but about a million FE dampers have been installed with the bolt; it's a robust fastener, and with adequate lube and thread engagement it will not get damaged. When in the correct finish location, the damper will be snugged up against the damper spacer. The snout of the crankshaft will not be flush with the

front of the damper; it will remain recessed a small amount. The damper bolt and damper washer are very strong, heavy-duty items. Thin or cheap fasteners will fail in this location. Original equipment–quality or ARP-racing-level hardware is required. Torque the damper bolt to specifications; you will either need a helper or to make up a fixture to prevent engine rotation while torquing this bolt.

The short-block is now completed. Let's flip the engine over on the stand and get the oiling system and pan installed.

OILING SYSTEM

The FE oiling system is very similar to the oiling system on other Ford engines. It seems to have a bad reputation among some, but the reality is that it is perfectly adequate for the majority of engine projects with minor modifications.

While Ford used a side-oiler system in the rare and valuable 427 engines after 1965, most FE engines will be center-oiler designs. That is the configuration we are going to concentrate on since it is going to apply to most folks using this book as a reference.

The oil pump is mounted to the driver's side front corner of the block inside the oil pan. It is driven by a 1/4-inch hex shaft that runs up to the distributor. Factory oil pumps were aluminum, but all the aftermarket replacement pumps are cast iron. The oil pickup tube and screen are bolted to the side of the pump.

You have the option of numerous replacement oil pumps for an FE build. I tend to gravitate toward two options: an M-57 standard-volume/standard-pressure pump or an M-57HV high-volume/standard-pressure pump. There is an M-57HP standard-volume/high-pressure version that is only for use with the factory 427 engines having a rear bypass (outside the context of this book, but good to know it is available). The pumps for the FT truck engines use a 5/16 drive.

The pump specifications are as follows (this information pertains to Melling brand oil pumps, and is courtesy of Precision Oil Pumps, a company that blueprints and customizes oil pumps for racing and high performance):

M-57 Standard-Volume Yellow
    Spring 40–45 psi 1/4-inch Drive
M-57B Standard-Volume Brown/Plain
    Spring 60–65 psi 1/4-inch Drive
M-57HP Standard-Volume Big Blue
    Spring 100–110 psi 1/4-inch Drive
M-57HV High-Volume Brown/Plain
    Spring 60–65 psi 1/4-inch Drive
M-57A Standard-Volume Brown/Plain
    Spring 60–65 psi 5/16-inch Drive
M-57AHV High-Volume Brown/Plain
    Spring 60–65 psi 5/16-inch Drive

Oil exits the pump into a short passage in the block. The block feed is one of the places we will modify. From the factory the inlet hole is very small compared to the outlet of the pump. We open that passage to match the pump outlet and smooth the flow

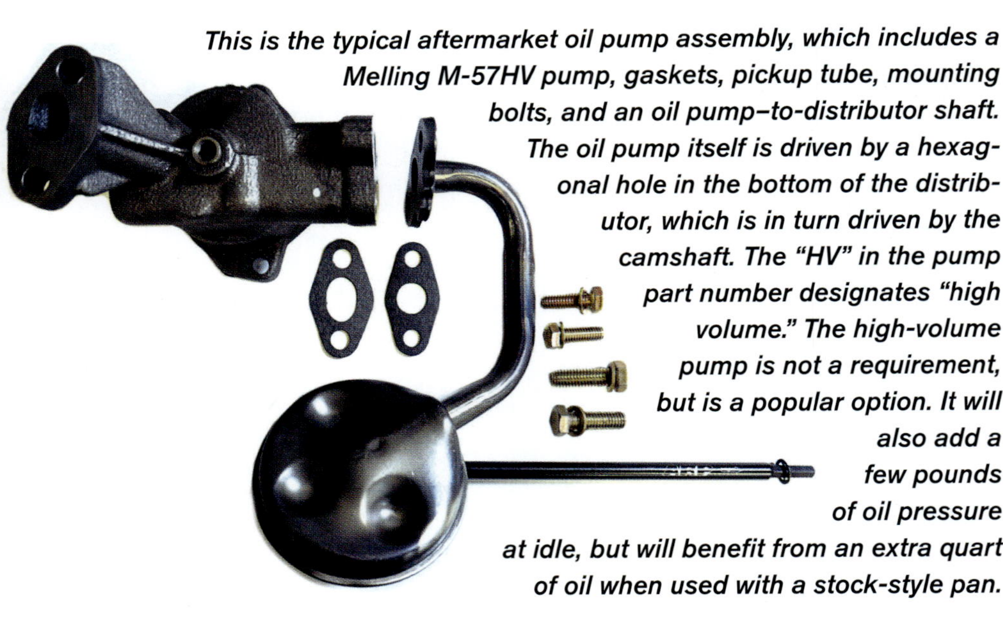

This is the typical aftermarket oil pump assembly, which includes a Melling M-57HV pump, gaskets, pickup tube, mounting bolts, and an oil pump–to-distributor shaft. The oil pump itself is driven by a hexagonal hole in the bottom of the distributor, which is in turn driven by the camshaft. The "HV" in the pump part number designates "high volume." The high-volume pump is not a requirement, but is a popular option. It will also add a few pounds of oil pressure at idle, but will benefit from an extra quart of oil when used with a stock-style pan.

The factory oil pump block feed is an area of restriction from the factory and one that is easily modified. Enlarging this feed to match the oil pump outlet will smooth the flow, improving it in the process.

Most replacement gaskets merely have holes for flow to the oil filter adapter. Since there is gasket material directly in the path of oil flow, a good policy is to trim away the extra gasket material. The gasket material over time gets brittle and can restrict oil flow by folding over or break loose and get into the system, clogging passages.

While many engines feed oil to the valvetrain via the lifters and hollow pushrods, the FE engine is an exception. Oil travels from the cam bearing, along the deck of the cylinder head to meet a head bolt hole, then vertical alongside the head bolt until it reaches a diagonal passage that meets one of the rocker mounting bolts. Factory rocker bolts in that location are usually necked down to aid oil flow up into the rocker shaft. This feed in the block must be clean and unrestricted, as well as crack-free. Any cracks or leaks will result in low oil pressure to the rockers and oil possibly getting into the cooling system.

path. The block passage routes oil out to the oil filter adapter. The filter adapter is an aluminum casting that bolts to the side of the block. There is a paper gasket between the adapter and block. This gasket is commonly supplied with punched holes to match the block openings. I prefer to modify it to match the adapter openings to try to prevent the paper from degrading, folding over, and blocking the flow.

After the filter, the oil returns to the block and runs up a diagonal passage in the block wall behind the timing set. This intersects and feeds a central front-to-rear passage in the lifter valley that serves as the main pathway for lubrication; thus the term "center oiler."

On most engines, seven passages feed off this center one. Five are vertical feeds that go to each cam bearing bore; two are diagonal to feed the lifters. The cam bearing bores each have an annular groove around them that allows oil to continue down to short vertical passages feeding the main bearings. The diagonal feeds intersect a pair of front-to-back feeds that supply oil to the lifters.

One unusual system design element on an FE engine is the oiling for the rocker system. Each cylinder bank has a passage in the block that goes from a cam bore to the cylinder head deck. The oil follows a convoluted path, traveling along the deck of the cylinder head to meet a head bolt hole, then vertical alongside the head bolt until it reaches a diagonal passage that meets one of the rocker mounting bolts. Factory rocker bolts in that location are usually necked down to aid oil flow up into the rocker shaft. Given this complex routing you would expect rocker oiling to be marginal, but the

All FE pans bolt to the block in a similar manner, but with external differences, depending on the application. This front sump pan is typical for Mustang and Fairlane applications, with indents at the rear that allow for tie-rod clearance. Some truck pans have center or rear sump designs, which clearly will not interchange into a typical passenger car application.

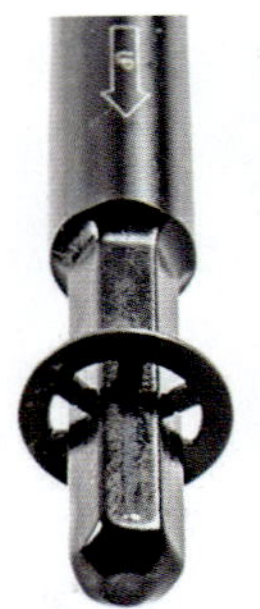

ARP makes an oil pump driveshaft that is considerably more robust than the factory offerings. Because of the retaining ring at the distributor end of the shaft, the shaft is installed at the same time as the pump. Never install this shaft without the retaining ring. The retaining ring is there to prevent the shaft from sticking in the distributor whenever you remove it. Often, the shaft only follows partway and then drops back into the oil pan. With the distributor mounting through the intake manifold, pulling the distributor occurs more often with FEs than with some other engine designs.

opposite is often the case. We frequently restrict the oiling by installing .060 or .070 inserts to the rocker passages to prevent flooding the valve covers. On most factory cylinder heads a Holley carburetor jet will do the job.

Remove the cover plate from a new oil pump and do a basic inspection to make certain there is no machining debris and to verify bypass valve operation. Reassemble the pump using a dab of Loctite on the bolts. Slip the driveshaft into position and then mount the gasket and the pump itself, again using the thread locking compound.

Oil Pans and Oil Pump Pickups

All FE oil pans are interchangeable to the blocks, but may be specific to their intended application. The majority of oil pans found on FE engines are going to be front sump configurations. The ones designed for Mustang and Fairlane applications will have a pair of indentations behind the sump for tie-rod clearance. Some trucks require oil pans with a center or rear sump. Engines using those will require that the long pickup tube be supported with a special main bolt or an additional fastener that goes into the block alongside the center mains. Make certain that you have provisions for that extra fastener if needed.

This Moroso 20607 piece is a T-type oil pan. The wings on the sides provide more oil capacity while not adding depth to the sump. Really helpful with lowered cars where ground clearance is important. They also offer them with extra baffles for road racing and high cornering forces.

A wide variety of aftermarket pans is available for FE applications, including the model 31130 from Milodon. This pan features a 7-quart capacity and baffling to reduce the risk of oil starving under heavy g loads. When using a high-capacity oil pan, always be sure to use the matching oil pump pickup tube. As in this case, most pan manufacturers also offer the pickup tubes that go with them.

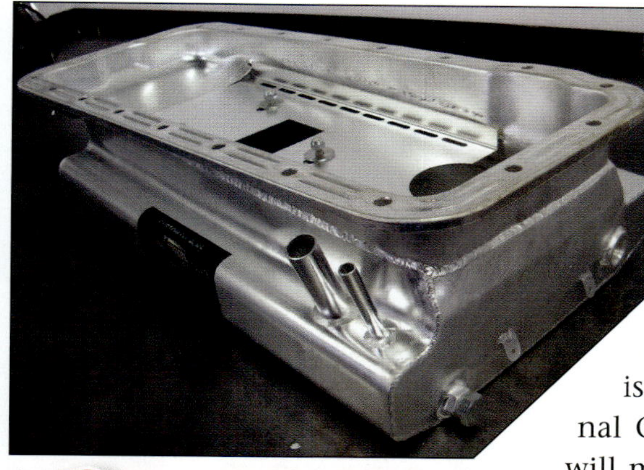

This is an Aviaid oil pan specifically designed for the 427 Cobra. Still available, this one is used by many kit builders for both its performance characteristics and its cosmetics.

With the oil pump mounted in place and the windage tray in position, we are now set to test the clearance between the oil pump pickup and the bottom of the oil pan.

a full-length iteration that is a reproduction of the original Cobra 427 item. Each of these will mandate the use of a matching pickup tube.

Next, we are going to install the oil pump pickup tube and check pan clearance. If you are using a windage tray it should be mounted to the block before installing the oil pump pickup tube. Clean the gasket surfaces with thinner or acetone; use just a dab of silicone at the junction of the timing cover and the block and at the rear where the main cap side seals meet the pan rails. Place an oil pan gasket on the block, and then the tray if using one. Lightly attach the pickup to the pump.

You can either measure from the bottom of the pickup to the pan rail and from the pan rail to the bottom of the sump for clearance, or simply put a ball of modeling clay into a sandwich bag and crush it between the pan and pickup. We want the pickup screen to be between 1/4 inch to 3/8 inch from the pan bottom. Too tight and you risk closing off the oil supply, too large and you risk uncovering the pickup during turns or acceleration. You can usually slot the mounting hole a bit or bend the pickup for minor adjustments.

Now you are finished with the bottom end. Flip the engine back over and get ready to start on the cylinder heads.

TECH TIP

Check Pickup Screen Clearance

Simply place a piece of modeling clay onto the pickup screen, secure the pan by tightening a few of the bolts, remove the pan, and measure the thickness of the clay. That is your clearance. On some pickups it's a good idea to put the clay in a plastic sandwich bag to keep it from getting squeezed into the screen. We want the pickup screen to be between 1/4 inch and 3/8 inch from the pan bottom. Too tight and you risk closing off the oil supply, too large and you risk uncovering the pickup during turns or acceleration.

A wide array of aftermarket oil pans is available for the FE. Among these are reproductions of the original parts, deep sump designs for drag racing, T-profile pans for road race and low ground clearance applications, rear sump specialty parts for rack and pinion conversions, and

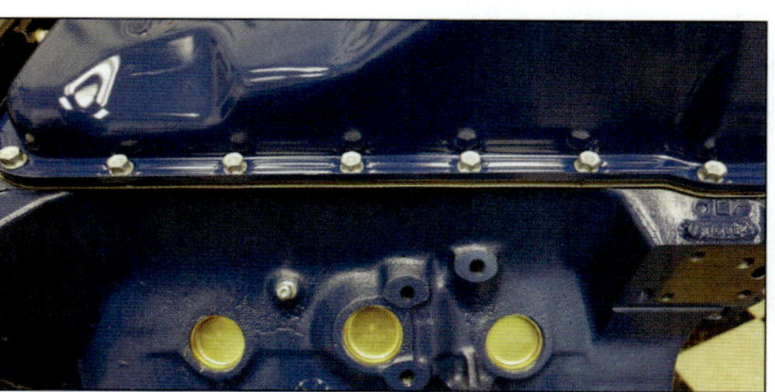

Once satisfied with the pickup clearance, make sure you install the pickup with a very light coat of sealer on the gasket and some thread locking fluid on the bolts for final assembly. Then set the pan gasket (you use two pan gaskets if using a windage tray, sandwiching the tray between them) and the oil pan in position. Pan bolts need to be lightly tightened in several steps to ensure that you do not deform the mounting surfaces of the pan. I prefer to use a serrated head bolt or lock washers since the pan bolts do tend to loosen over time.

CYLINDER HEADS AND VALVES

Cylinder heads on all FE Ford engines share the same general architecture. They all have the same 10-bolt pattern for mounting to the block, similar valve cover mounting, and the FE-unique split at the intake manifold where the pushrods go through the manifold instead of the heads like on other V-8 engines. Several design elements have changed during the 20-year span of production, and some of these will have a definite impact on whether a particular set can be used for your project. The most relevant factor is the exhaust port position, which affects header compatibility.

Basic 360 and 390 Heads

Most FE builders follow one of three distinct paths when choosing cylinder heads. If you are restoring a vehicle, you need the exact heads that were used originally. If you are building for performance, you will likely gravitate toward the casting that provides the most power for your money, and you may end up going to an aftermarket product. If you are building just to get a nice running engine for your personal project, you will choose what you can find and afford to get the job done.

Most FE cylinder heads are going to be standard parts from a 360- or 390-powered pickup truck or a 390 passenger car. These share the same exhaust bolt pattern with eight bolts vertically oriented in pairs above and below each port. All traditional eight-bolt FE exhaust manifolds and headers will physically mount onto any of these heads. But the port positions changed in later years. They were lowered relative to the bolt pattern, and they will not interchange without risking significant port restriction and leakage. Ford FE heads usually have a casting part number located in between the center exhaust ports on the outside of the casting. The casting number will help determine the age and application of the heads.

If you are working on a later (after the mid-1960s) pickup truck, make sure you have the correct headers before swapping out heads. If you already have the exhaust, you must stick with the matching heads. Header availability will often drive the casting decision.

The D2TE-AA 1972 and later castings have hardened exhaust seats from the factory and might be a touch more tolerant of long-term use

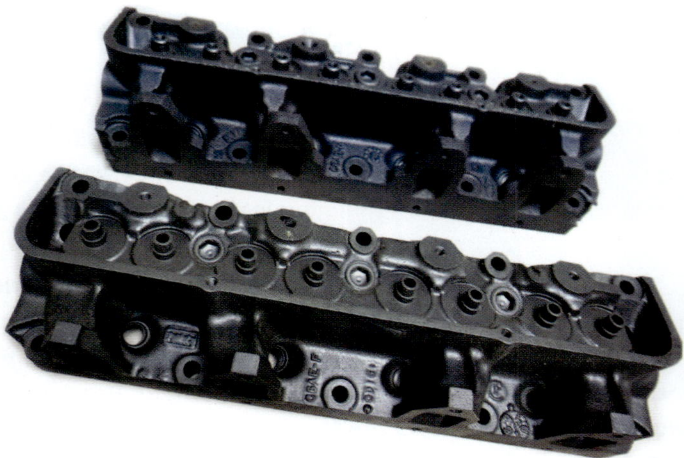

Fortunately for FE enthusiasts, there is very little variety in the architecture of the heads. They all have the same bolt pattern for mounting on the block, as well as valve cover mounting. The most important running changes were in the exhaust mounting bolt patterns, as Ford had to make allowances for the tighter engine bays of intermediate models.

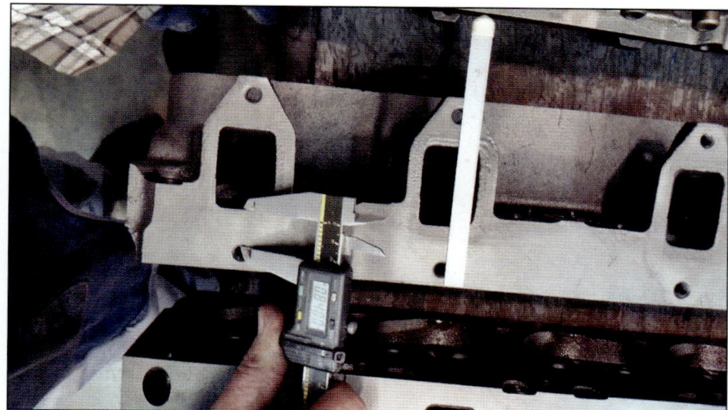

Not only did some of the bolt patterns vary on the exhaust side over the length of the production run, the distance between the bolt holes and the exhaust ports did as well, an important consideration when fitting different exhaust manifolds or aftermarket headers.

Later heads, such as this example, are +/- .800 as measured from the top of the port to the bottom of the upper bolt hole, and would be considered a low-exit port. They helped with shock tower clearance but were not particularly good for performance.

This illustrates the potential port mismatch between the low-exit port head shown and the high-exit port variation shown in the image to the right. You can physically bolt the headers on to either head, but this kind of mismatch can go undetected and will result in continuous header leaks, along with reduced performance.

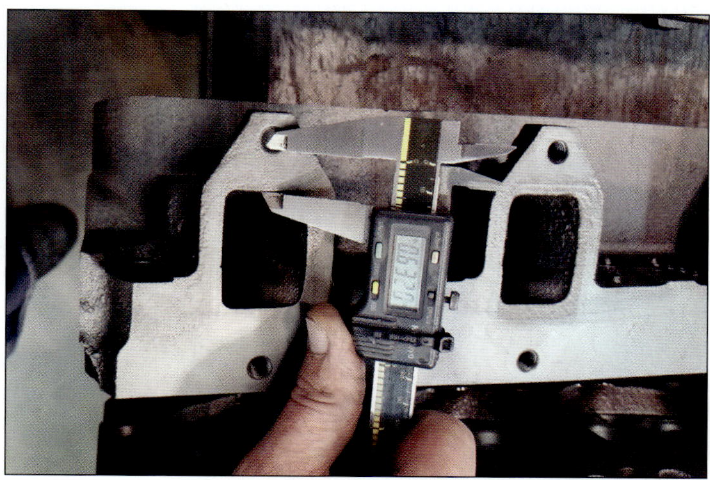

This earlier, higher-exit port head is .632 inch as measured from the top of the port to the upper bolt hole. Better for performance, these will still clear shock towers with headers, but cast-iron factory manifolds will not.

with unleaded fuels. They also have a rather small intake port that favors low-RPM torque and response. They were designed for truck use and perform well in that application.

Same thing goes in the opposite direction for early Galaxies or Thunderbirds. Earlier castings often have a larger intake port opening with a lowered floor that rises as it goes farther into the head. I suspect that this low entry was done to accommodate the intake manifolds required to clear really low hood lines that were coming into vogue in that era.

There may be a performance advantage to the larger opening, but it is modest when using factory heads in a street application, especially if any of the heads get upgraded with an enhanced valve job and bowl work.

The 390 GT heads are unique to the Mustang and Fairlane platforms. They have a 14-bolt exhaust pattern with the outermost upper holes in a different spot than in other FE engines. The heads themselves are nothing special in terms of performance potential and are not considered an upgrade. The headers designed for these will not fit other heads, and these heads cannot easily be used in other vehicles as a result. Probably best to leave these for the folks restoring the cars they originally came on.

The 428 Cobra Jet heads are unique to the 1968–1970 cars built

The 390GT heads are somewhat unique in that they have a 14-bolt exhaust pattern with the outermost holes in a different location from other FE heads. These heads are unique to Mustang and Fairlane body styles to accommodate the narrower shock towers and are best left to numbers-matching restorations. The headers for them are unique to them, and trying to interchange with other applications is more hassle than benefit.

The 428 Cobra Jet heads are also unique in that they have a 16-bolt exhaust pattern, and these heads do offer a performance benefit over standard FE heads. Cobra Jet heads are among the better performing factory castings and come with larger 2.09-inch-diameter intake valves and 1.65-inch exhaust valves.

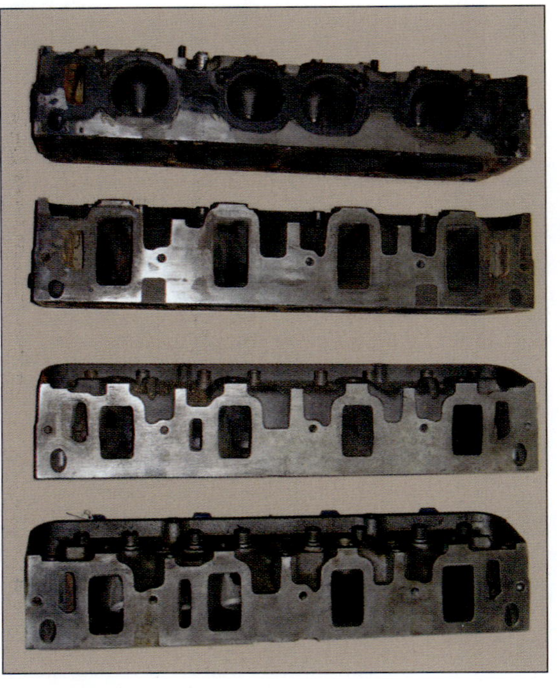

This is a nice illustration of the various FE intake faces over the years. The top one is a tunnel port. Odds are you won't find any of these just sitting around, but they are cool to look at. The second one down is a 427 high-riser. Could even be less common than the tunnel ports but served as the heart of the engine that powered the famed Thunderbolt Fairlanes of 1964. The bottom two are much more common. Although the actual head shown as the third head down is a valuable 427 medium-riser, similar size and location port openings are found on other OEM FE heads, as well as on the popular aftermarket heads from Edelbrock. These would use the Fel-Pro Performance 1247-S3 intake gasket.

The bottom head port has a low-riser–style opening and was used in numerous factory FE engines ranging from 427s to normal 390 and 428 applications. The larger opening calls for a 1246-S3 Fel-Pro Performance gasket or the common MS90145 passenger car variant.

with that engine. They are a nice performance upgrade over common FE heads and all share the C8OE-N casting number. Cobra Jet heads have a unique 16-bolt exhaust pattern. The normal 8 bolts are in the standard position, and an additional row of bolt holes are in a staggered horizontal pattern. Cobra Jet heads are among the better performing factory castings and come with larger 2.09-inch-diameter intake valves and 1.65-inch exhaust valves. These larger valves can be installed into other castings as an upgrade.

The 427 heads came in a wide array of designs over the six-year run of vehicle production. A few of these were very low-production high-performance specialty items and the odds are extremely high against you stumbling across a set of them in any junkyard, swap meet, or parts vehicle. The 427 tunnel port or high riser heads were rare factory

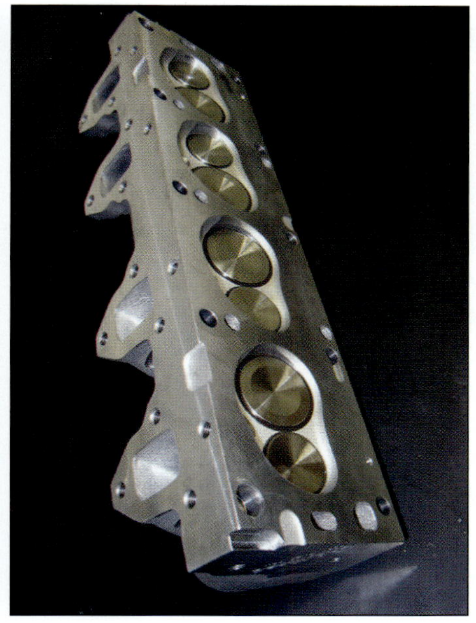

Edelbrock is the most common head manufacturer for the FE engine, just as they are with many other of the less-than-mainstream engine designs. Edelbrock has been the most aggressive at providing heads to the market where previously nothing else was available except tired old cores needing extensive work from the salvage yard.

Edelbrock's Performer RPM 72-cc head mimics the 16-bolt exhaust pattern found on the factory 1968–1970 Cobra Jet heads. The intake ports are similar to the 427 medium-risers, making this head a bit of a hybrid design in comparison to the factory offerings. Minor modifications in design have been made to ensure these heads will fit all 390, 427, and 428 engines.

performance items and fall outside the context of this book; the 427 medium-riser heads are somewhat more popular but still quite rare. The 427 low-riser heads are the most common, generally similar to normal FE castings, and still not something seen in the average build.

Aftermarket Head Designs

By far the most common aftermarket cylinder head for an FE engine build is the Edelbrock. The company uses a single casting and generates multiple finished heads from it, with changes limited to combustion chamber machining and valve springs matched to camshaft types.

The Edelbrock heads are something of a hybrid design, with a similar exhaust to the 428 Cobra Jet, and an intake that closely matches that of the 427 medium-risers. The exhaust valve moved slightly to arrive at a part that will physically mount to any 390, 427, or 428 engine. You still need to check for header fit related to the port position.

All Edelbrock heads (except the semi-finished Pro-Port iteration) use the same basic casting and alter it to meet various requirements. The 72-cc chamber version has a contoured "as cast" chamber, along with the 428 CJ 16-bolt exhaust mounting pattern. The 76-cc version uses a flat machined chamber and has only the 8-bolt 427 exhaust pattern. Intake ports are all 427 medium-riser–size opening and position and fit to the Fel-Pro 1247 gasket. Valvetrain compatibility with most common aftermarket and factory 390 and 428 systems is possible, although the relocated exhaust valve may require a bit of work with shims to properly line up the valve tip with the rockers.

The Edelbrock head offers a modest power increase over a factory 428 CJ or 427 medium-riser head, and a more significant one over standard 390 heads. With a retail cost of under $1,800 per pair, they become a reasonable alternative to factory heads that require major repair.

Survival Motorsports offers an improved high-performance head that shares most of the compatibility with the Edelbrock parts but includes significantly altered ports and an updated chamber for enhanced performance. The Survival heads cost roughly $2,500 per pair and are designed to support more than 500 hp. The intake port is wider than stock and will require some port match work on the intake manifold to get the best results. These are a great choice for a higher-performance street application, but only you can decide if the potential power gain justifies the added cost on a milder build.

This is a Survival Motorsports aftermarket FE aluminum head. It is designed to fit the same applications as the Edelbrock in terms of manifolds and headers, but provides a significant power increase for folks who are making a serious upgrade. It includes reconfigured intake and exhaust ports, along with a more modern heart-shaped combustion chamber.

All the threaded bolt holes will need to be clean and square. Damage often occurs to the threads through exposure to the elements, incorrect installation techniques, or improper-length fasteners. Often a Heli-Coil, as shown here, is the best cure.

Cylinder Head Reconditioning

For the engine being covered in this book we made the decision to rework the original heads. This work is generally done at a machine shop, rather than at home. With FE engines now approaching 50 years of service, it would be rare to find a pair of heads that can be effectively reworked by simple cleaning and valve lapping. We are detailing the machining and assembly process in the following section. If you choose to simply replace your heads, it may still be useful to follow along, checking your new heads for things like spring tension and assembly clearances.

Castings: Cleaning, Inspection, and Qualification

The first step with your chosen heads is similar to that taken with the engine block. They are carefully inspected for damage and then thoroughly cleaned. Factory Ford FE heads are subject to cracks in certain locations, damage to threads, and fretting around the head bolt bosses. Some of these issues are a reason for discarding and starting anew, others are repairable or not a cause for major concern. Given the low cost of replacement castings, it is usually not worth investing a lot of money in repairing normal passenger car or truck heads, unless you are working on originality for restoration purposes. The 428 Cobra Jet heads, such as the ones on this build, are very valuable and justify extra effort and expense.

We sometimes see cracks in the factory heads. The fix or replace decision is one of value determination; I will list some likely locations to look. Freezing coolant will crack them horizontally on the external wall below the valve cover and between the center exhaust ports. Occasionally we find cracks that originate at an intake valve seat and propagate across to the spark plug hole. These can sometimes be functionally fixed by installing replacement hardened valve seats.

Bolt holes are very often damaged through use of improper-length fasteners, poor installation techniques, or exposure to the elements. All of the threaded holes will need to be chased with a tap, and numerous Heli-Coil inserts are often installed. You can certainly install a thread insert at home, but the need for each threaded hole to be perfectly straight and in line with its companion holes means that it's a job best done in a machine shop.

The sealing surfaces of the cylinder head are all going to need attention. Cylinder head gasket sealing is highly dependent upon having a clean, flat, and smooth surface. While a budget-oriented project might get by with a few swipes from a sharp, flat file, sealing integrity at the head gasket is critical to a successful build. Machining this surface is one of those operations that should be considered a high priority in any quality rebuild effort.

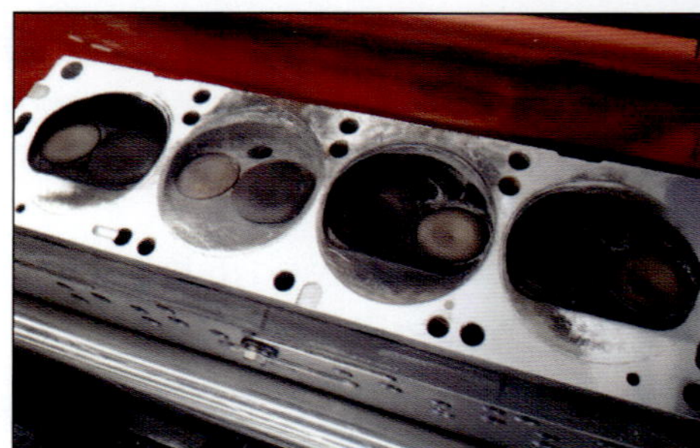

Similar to the block deck, the sealing surfaces of the cylinder head will certainly need attention. It is vital for proper head gasket sealing that both the block deck and cylinder head deck are clean, absolutely flat, and smooth. This is a step, as are most steps in cylinder head reconditioning, best left to a quality machine shop.

The exhaust face and intake face should also be smooth and flat, and can be lightly dressed with a sharp, flat file on a budget build. It is considerably better to have these surfaces machined using the same machine that will be used to surface the cylinder head gasket deck. Considering that intake and exhaust gasket leaks are among the most common complaints with FE engines, this remains another place where the investment in machining will pay off.

Component Selection

Although head components are normally installed by the machine shop, we will run through the selection options.

Valve Guides

Factory heads all came with integral cast-iron valve guides; they simply machined the iron castings for correct guide dimensions and clearances. FE engines came from Ford with 3/8-diameter valve guides. Correct valve guide clearances are very important for good sealing and oil control. Valve stem–to–guide clearance recommendations will usually be somewhere between .0015 and .0020 of an inch, depending on guide material and anticipated engine use.

Valve guides tend to wear out over time and are serviced in a few different ways. Valve guide repair or replacement is considered a machine shop operation, something you will not be doing at home. Some repair techniques have fallen out of fashion as the cost of small parts becomes less important than the quality of the resulting repaired item.

Guides can be final sized by either reaming or honing. Honing is considered a better finish process for use in guides, but plenty of shops still rely upon a reamed finish. The least expensive guide repair is to use a slightly oversized valve. The existing guide is then reamed or honed to a larger diameter to get the proper clearances. This was common practice with large volume commercial engine rebuilder shops, but you will not find many folks doing this

A few methods have been developed to properly recondition the valve guides in your original castings. Original guides were often merely a 3/8-diameter guide machined directly into the cast-iron head. Often the repair for the guides was to simply enlarge them and get valves with oversized valve stems to match the new diameter. This practice has fallen out of favor, especially in engines like the FE, where valve selection has become more challenging.

Another option is to knurl the guides, which involves a special roller that forms a threadlike pattern in the guide, forcing iron material up so that clearance is returned to factory specification. This has fallen a bit out of favor as well, especially in more performance-oriented shops, largely because the repair is fairly short-lived. As you might guess, with the reduced surface contact area in the threaded guide, wear occurs more quickly.

The best guide repair is the use of an insert or full replacement guide installed into the enlarged previous guide. The most popular material for the inserts is bronze, which has better wear and sealing characteristics. Cast-iron inserts are also available, but most people choose the bronze inserts.

anymore. Oversized valves are pretty difficult to find for FE engine builds.

The next option is the use of a small roller on a mandrel that is inserted into the existing guide and turned to "knurl" the guide. The roller forms a threadlike pattern as it travels down the guide and raises the adjoining surface to tighten up clearances. This was considered an effective low-cost solution to tighten up loose guide clearances. It is a short-lived repair, though, and is no longer a preferred method in performance-oriented machine shops.

The use of an insert is generally regarded as the best repair for worn valve guides. Inserts can be either a thin-wall design or a larger-diameter replacement guide type. The thin-wall version is a bronze sleeve that is pressed into the existing guide after it has been reamed to a larger diameter. The sleeve is then honed for clearance.

The full replacement guide is a larger-diameter insert that completely replaces the integral guides in the head. The original guides are drilled out to a much larger diameter, which completely removes the original part. With an outside diameter of around .502, these are intended to press into a .500 hole. The new guide can be made from either cast iron or bronze. Iron is an excellent guide material, but bronze is by far the most popular choice in performance applications. Once installed, the new guide is honed for clearance.

Valve Seats

The valve seat is the part of the head that the valves seal against. It should be machined for the correct angles for sealing and airflow, and be smooth and concentric with the valve guide. Sounds pretty sim-

The shiny ring around each valve seat is where it contacts the valve. These are showing pretty good contact and a functional seal.

The dark spots around the exhaust seat tell us we need to do a valve job; things are not seating and sealing as they should. This example shows normal wear for an older engine.

ple, but a quality valve job can have a profound impact on engine performance.

A factory FE valve seat is integral to the head; it is simply machined into the iron casting. Heads dating from the 1970s on often have factory seats that are induction hardened to improve durability with unleaded fuels. Given that any factory FE cylinder heads are now more than 40 years old, it is safe to assume that they will either need work or that they have already been worked on. Possibly several times.

Evaluating the valve seat's condition starts out with a visual inspection. The desired appearance is that of an uninterrupted bright and clean contact line around the entire

diameter with no pits or discolored areas. Unless the heads have been freshened up recently, you won't often see this good of a pattern on the intakes, and almost never on the exhausts.

Most often the valve seat on any older heads needs true reconditioning. It will always need rework to re-establish concentricity if the guides have been serviced. The valve seat has several critical characteristics. As previously mentioned, it needs to be concentric to the guide so that the contact with the valve is perfect all the way around. The contact angle must be correct for both valve and seat, and the contact width must be adequate to provide long-term durability.

Concentricity of the valve seat is important because any degree of off angle or partial contact will cause uneven loading of the valve. This in turn will rapidly wear out guides and seats, and invite catastrophic valve failure in extreme cases. Concentricity is measured with an indicator mounted to a stand inserted into the valve guide. If the seats are beat up or the guides are being replaced you won't need to worry about doing this check now. You can simply assume that it is going to be bad, and that it will be addressed during the valve job. A valve job is mandatory following any guide repair.

A valve seat that is badly worn or sunk into the head needs replacement. This entails installing a seat insert into the head, which is definitely a shop operation. The replacement seat will be a ring that gets pressed into the head and subsequently machined. To accommodate a replacement seat, the head must be machined, a delicate operation in FE heads. Once a receiving pocket has been machined into the head, the new seat is pressed into place.

Valves

The valves used in the majority of factory FE engines are similar in basic design. They are made from a high-quality steel alloy, a stainless steel on the exhaust side. They have a nominal 3/8-inch-diameter valve

Here is a worn cylinder head before machining. You can clearly see how far the exhaust valves are sunken into the chamber. This is an extreme example but shows the result of long, hard usage with unleaded fuels on a nonhardened seating surface. New hardened seats that are compatible with today's fuels are a good option to consider for any cylinder head rebuild using older castings, and are mandatory as a service replacement for damaged castings.

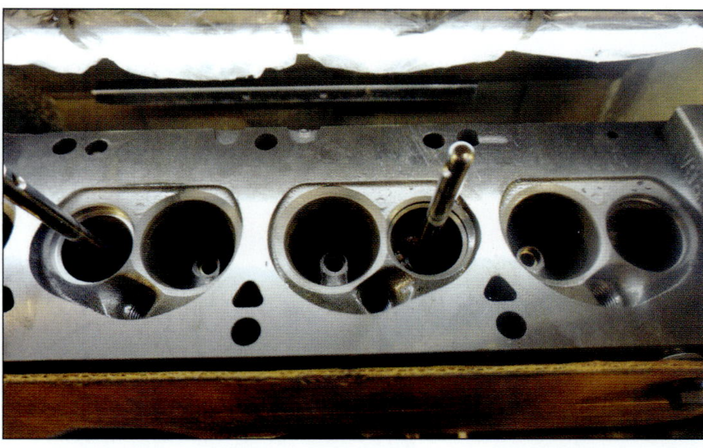

Shown here is a factory iron head being prepped for hardened replacement valve seats. Toward the center you can see an exhaust seat that has been severely degraded in service and which will require replacement. To the far left is a seat that has been machined to accept the new replacement insert.

Since the factory seats were integral to the head in FE engines, installing hardened seats will require machining a recess in the casting for the replacement seat insert to be pressed into. The new seat, once pressed in place, is subsequently machined as traditional valve seats are. Here you can see the replacement seat insert just laid in position above the intended location.

This image shows the new replacement seat ring as installed in the head before the valve job is done.

These are typical valves for the FE engine, made of steel alloy and stainless steel on the exhaust side. Most standard FE valve sizes are 2.03 inches on the intake side and 1.56 inches on the exhaust. The valves in our Cobra Jet head are larger in diameter, though, at 2.09 inches and 1.65 inches respectively.

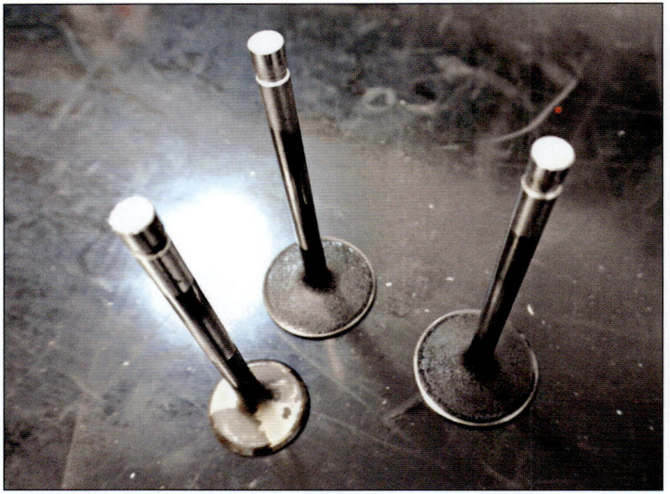

Here we can see the used valves. The bright ring around each intake valve head is the seat, just like those on the cylinder head itself. The deposits on the back side of the larger intake valves are indicative of oil leakage, most often from worn-out valve guides and valve stem seals.

This angle shows the valves where sealing was compromised. Pitting or interrupted wear at the seat area show us that the valves were not sealing and performance was affected.

The flat surface around the outside diameter of a valve is referred to as the "margin." A minimum of +/- .060 in margin thickness must be retained to avoid a sharp edge, which would be prone to fracture or creation of a hot spot in the chamber. If a worn valve requires so much grinding that the margin is eliminated, it is considered junk and needs to be discarded.

Valve Inspection

A basic inspection and measurement will be necessary when deciding whether to recondition or replace the valves in your FE heads. The decision to replace or discard is often an easy one, as new valves are fairly inexpensive. Anything showing less than near perfect condition will likely get replaced.

Mounting a valve in a lathe, valve grinder, or even a drill will quickly identify a bent valve, which is not usually worth reworking. Any valve with obvious physical damage (chips, scratches, cracks) will get thrown out. Same goes for a used valve with significant wear on the tip or stem. The angled surface that contacts the head should not show deep wear or excessive pitting. This area gets ground during refinishing, but there are limits to how much material can be removed.

stem that will always measure a few thousandths smaller than that; the desired stem-to-guide clearance is in the valve, so stems will be +/- .373 in diameter. They have a single groove for the valve keepers and a hardened top at the rocker arm contact end. The most popular head diameters will be 2.03 inches on the intake and 1.56 inches on the exhaust. Some performance applications, such as the 428 CJ, use larger 2.09-inch and 1.65-inch valves, which can serve as an upgrade to standard heads. The seat contact angle will be either 30 or 45 degrees on the intake and always 45 degrees on the exhaust.

The Valve Job

The term "valve job" is used as a general description of the cylinder head reconditioning process. The key part of that task involves the grinding of valves and valve seats to attain a good seal, after all of the cleaning and inspection previously described. If you are on an extreme budget, and the existing contact surface and pattern looks pretty good, you could conceivably just touch up the valves and seats with some lapping compound. This is an abrasive paste material. You put a small amount between the valve and its seat, attach a suction cup on a stick to the valve, and spin it back and forth with your hands. The abrasive will remove some embedded dirt and often makes it easier to see the underlying seat condition. Make certain to thoroughly clean the surfaces and remove any abrasive afterward. Brutally simple, decidedly old school, and don't expect top-quality results. But better than doing nothing. We sometimes use this technique to inspect and verify seat placement and angles on a new valve job.

Valve seat reconditioning is a task best performed in a professional machine shop. It was common to use a grinding process with abrasive stones that established the desired angles. These days it has become more common to machine the seats with carbide cutters on a dedicated seat and guide machine. Either method will give good results in the hands of a skilled operator. The seat and guide machine is both faster and more accurate in most cases, while stones allow for a degree of creativity and flexibility that is tough to duplicate. Most shops use the machine these days because it can also be used for guide and seat replacement.

This is a Sunnen VGS-20 valve machine. Most professional shops use a similar piece of equipment for valve seat and guide work.

Here you can see we are starting in on machining the replacement valve seat that was installed earlier in the chapter.

The valve machine uses a specialized cutter to form the angles desired for a high-quality valve job. Valve seat angles in the head are specified to seal to the valve, along with additional angles just above and below the seat that help direct and enhance airflow. Finished valve seats must be accurate in angle, as well as being smooth surfaced and concentric to the guide.

FE cylinder heads always use a 45-degree seat angle on the exhaust valves, and may have either a 30-degree or a 45-degree seat angle on the intake valves. While it is possible to alter the seat angle for certain performance applications it is far more common on street-oriented builds to simply use the seat angle that matches that of the original or replacement valves.

Performance builds usually use additional angles on the valve bowl and combustion chamber sides of the main sealing angle. The idea behind the additional angles is to help promote better airflow from the port

This machine is used to resurface the valves themselves. One grinding wheel refinishes the seat angle. With valves that are being reconditioned, it is always a good idea to clean up the tips of the valves as well where they come in contact with the rocker arms. Be careful about how much material is removed; the hardened surface is quite thin.

Valve springs and keepers, considered wear items, should be replaced with any rebuild. Even though they may pass a visual inspection, after millions of cycles of compressing and contracting, they are fatigued to the point where performance may be compromised. Only the most budget-oriented builds should consider reusing the springs after inspection.

After reworking both the valves and the seats, we can check the new sealing surface by coloring them with a Sharpie marker and rotating them against each other. The marker will clearly show the position and width of the new sealing surface.

As with the marking on the valve, coloring the head's seat will indicate the position and location of the new sealing surface.

into the cylinder. The enhanced flow makes more power and is a worthwhile upgrade even on mild engines, at a fairly modest cost.

A good valve job will be concentric with the guide, having sharp, well-defined angles, with smooth seats free of chatter. Done well, all the seats will be at a similar height to each other as referenced from the deck surface.

The valves themselves are ground to correct angles using a dedicated grinding machine. Used valves are always ground, and it is prudent to at least touch up new ones, just to be certain. The same machine can be used to detail the tips of the valves where the rockers contact them. You

do not want to remove much material from the tips since they have a thin hardened surface.

Valve Springs, Locators, Retainers, and Keepers

After you have sorted out the valves and valve seats, it's time to look at the rest of the hardware that comprises the valvetrain. All components need to be in top shape to work well with each other and deliver the best performance.

Valve Springs

In most builds the valve springs should be replaced. They are wear

items and are crucial to good performance. All common factory FE engines used a single spring with a flat wire damper to control harmonics. Many of them used a doubled-up retainer that was designed to allow for valve rotation, whereas the high-performance versions used a stronger one-piece steel retainer. Retainers need to match the chosen valve springs, but the original parts are almost always suitable for reuse on lower-budget rebuilds. We address each of these parts in the following paragraphs, but remember that they work together as a system.

Spring Sizes and Pressures

Valve springs have several critical dimensions that must be considered during selection and installation. These involve both installation and operational characteristics. Although an easy statement to make, it's always best to at least reference the spring recommendations from your cam manufacturer. Going too far away from those values will sometimes cause problems that you do not want to face.

Our first concern is spring pressure. Valve springs are rated for installed pressure, the measured pressure in pounds when the valve is at rest in the closed position. They are also rated for the pressure when the valves are fully open, and for the rate of pressure increase per inch of travel. Valve spring pressure measurement requires an expensive dedicated tool, and is therefore a machine shop process. If you are working at home your best bet is to measure the installed height, select new springs based on application and manufacturer's specifications, and hope for the best.

Springs intended for use with flat tappet cams, whether solid or hydraulic, have lower installed and open pressures than those intended for use with a hydraulic roller camshaft. Flat tappet cam springs most often have installed pressures between 110 and

Many manufacturers use dual valve springs to increase spring rate in performance applications. For the FE engines, we often go to a double spring on applications with higher-RPM intentions and solid flat-tappet or hydraulic roller camshafts.

130 pounds coupled with open pressures between 280 and 350 pounds Too much pressure on a flat tappet cam will cause premature wear and possible failure of the lobes. If running double valve springs, we always remove the inner spring during cam break-in procedure as insurance against early problems. Just remember to install the inners before going to high RPM once break-in is finished.

A hydraulic roller cam allows and requires higher pressures to function well. We often run between 140 and 160 pounds of seat pressure and around 375 to 425 pounds of open pressure on those. The roller cams eliminate the break-in concerns, and the heavier lifters require more pressure to keep the valvetrain under control at higher RPM.

Solid roller cams are outside the target engine of this book, but as a reference they run far higher valve spring pressures. They start out at more than 200 pounds seat pressure and more than 500 pounds open, with some going beyond 300 pounds on the seat and 1,000 pounds open. These pressures require highly modified rocker assemblies and frequent inspection and replacement. Solid rollers are race car stuff for the most part and fall outside the context of this book.

Outside Diameter

The next critical dimension is the outside diameter. A larger-diameter spring can be stronger, delivering higher pressure and rate while avoiding coil bind. The FE head is a forgiving design in that it allows for a 1.50- to 1.55-diameter spring with no modification, giving us lots of spring choices to work with. Just be certain that both the retainer and locator

match the chosen spring. If you are running a double spring, make sure to account for the additional locating step in the retainer when calculating and measuring spring pressures.

Most FE builds use either a single spring with a flat wire damper or a double spring. The double spring will not usually require a damper, although some still have them. Single springs are most often used on stock or near stock builds with flat tappet cams. Higher-performance flat tappet cams and most hydraulic roller cams are better served with a double spring.

Beehive and conical valve springs have recently become popular in the performance aftermarket. They both feature a spring winding design that gets smaller in diameter as it goes up from the head's surface. This allows use of a very small and lightweight retainer, and the characteristics of the winding shape reduce harmonics so that a damper is not required. The weight reduction allows a few hundred additional RPM before valve float occurs. If you choose to use them, you will absolutely need to acquire a matching locator and retainer from the manufacturer.

Coil Bind

Coil bind is the point where the valve spring goes solid and cannot be compressed any farther. While rarely a problem with stock or near stock rebuilds, this can become a real issue on higher-performance combinations running high-lift cams and is a reason you should go to a double spring on those. The thinner wire used on a double spring allows more lift before binding up, while the addition of the smaller inner spring brings both the installed and open pressures up to a higher level suitable for high-RPM use. It is generally good practice to provide at least .060 inch of clearance beyond the coil bind point at maximum valve lift. Having more will never hurt you on mild street builds, but having less will cause a great deal of trouble, breaking expensive parts.

Before leaving the discussion about bind and moving on with assembly, I should mention that other parts here can get into a mechanical bind at full valve lift. Retainers can hit the top of a valve guide or valve seal, and rocker arms can hit the sides of a retainer if larger-diameter ones have been installed. Pay close attention to such things during assembly, catching and correcting for them now will save a lot of grief down the road.

Measuring and Equalizing Installed Height

Installed height is the measured distance between the top and bottom of the valve spring in its closed position. Installed height will have a major impact on spring selection and must be at least roughly established before moving much further on any project.

This can be measured with a variety of tools depending on your need for accuracy, time, and budget. A production shop will use a dedicated height micrometer that is installed between the cylinder head (or spring locator) and the retainer. The keepers are installed into place, the retainer is pulled up to its finish location, and the thimble of the height mic is rotated until snug, providing a direct and accurate reading on the installed height. Someone working at home can get a reasonable measurement with a machinist's ruler and a straight edge, or a snap gauge and dial caliper. For most builds accuracy within

You should always measure the installed spring height of your valve springs, which is the difference between the top and the bottom of the spring while the valve is in the closed position. This spring height micrometer does an excellent job of measuring the distance, although it is a tool you will find mostly in professional shops.

.010 to .015 is plenty good enough.

Installed height is adjusted with spring shims that are placed under the spring or under the locator, if used. Spring shims are most often available in thicknesses of .015, .030, and .060 inch. The need for more than a couple of shims is usually an indication that a different valve spring should be considered. Specialty retainers and keepers are available to make significant changes in installed heights for high-performance applications. These are very useful, but don't plan on using both the +.050 retainers and the +.050 keepers at the same time. You can get to a point where the rocker arm will contact the retainer before it contacts the valve, which is a recipe for broken parts.

Locators

All valve springs need to be pos-

Installed height is adjusted with spring shims, which are placed under the spring, or under the locator if used. Springs shims are most often available in thicknesses of .015, .030, and .060.

itively located both at the top and the bottom. The factory cylinder heads have a raised diameter around the valve guide that serves as a locator. If that area has been machined flat during guide replacement, or to accommodate aftermarket springs, a separate locator or spring cup must be used. Aluminum heads always require a steel locator to protect from

The factory cylinder heads have a raised diameter around the valve guide that serves as a locator. If that area has been machined flat during guide replacement, or to accommodate aftermarket springs, a separate locator or spring cup will need to be used.

abrasion and wear. All spring locators have an inside diameter that closely fits around the valve guide. Some aftermarket locators have a step that indexes to the inside of the valve spring to keep it in the proper position. Others will take the form of a spring cup that has a raised outer diameter to hold the spring in place. Most spring locators will be .060 inch in thickness; this dimension needs to be included in spring height calculations.

Original FE retainers are often two-piece assemblies that were designed to promote valve rotation in extended low-speed service. While usable in stock replacement applications, these are most often replaced in any modern build with an eye toward performance and durability.

Replacement retainers for higher-performance use are machined from high-strength 4130 chrome-moly steel. They feature a step underhead to index into and positively locate the desired valve springs.

Retainers

Valve spring retainers have a locating step, or multiple steps for double springs, to index and retain the upper valve spring. The valve spring must be well supported around the outer diameter and be a reasonably snug fit around the inside diameter. A press or snap-tight fit is not required, but they should not be loose or sloppy.

The majority of retainers, both OEM and aftermarket, are machined steel. Some factory stuff is stamped steel. Many original retainers are a two-piece design that was intended to promote valve rotation, which was thought to help keep the seat contact area free of carbon buildup. Factory high-performance retainers and all aftermarket retainers are a stronger and lighter one-piece configuration.

Most aftermarket replacement retainers will be higher-quality chrome-moly steel, which is suitable for all but the highest-performance applications. Alternate lightweight materials such as tool steel or titanium are offered for high-RPM racing use, but they generally fall outside the spectrum for the type of engine we are building in this book. The only notable issue of concern is the use of aluminum retainers. They were somewhat popular back in the 1970s, but they have proven unreliable and if found should be discarded and not used in any build.

Keepers

All factory FE retainers are designed to work with single groove 7-degree keepers (also referred to as locks). Although the factory keepers are often perfectly fine, the replacement cost is so low that replacement is advised in most cases. Aftermarket retainers and keepers may have

While replacement isn't absolutely necessary, the cost of replacement valve spring keepers (also known as valve locks) is so low that you really should replace them with every rebuild. They are cheap insurance when you consider what could happen if they fail. A dropped valve and a holed piston are not worth the risk just to save a few dollars on keepers. When using aftermarket retainers, always stick with one brand and angle to ensure proper fit.

Any retainer and spring combination should have a snug, but not excessively tight, fit between the chosen parts. Retainers that fit loosely in the spring will not function well, allowing the spring to wobble and bounce around at higher RPM.

different angles and diameters, so stick to a single brand when choosing these parts to be certain that they will fit properly together. A 7- or 10-degree keeper from one brand is not always compatible with the retainer from another brand, even though the angles are the same. Some manufacturers offer optional keepers that allow adjustment of spring installed heights, a feature that can come in handy during head assembly.

Setting and Inspecting Stem Tip Height

Most factory FE engines have a nonadjustable valvetrain. We therefore rely on the accuracy and consistency of the machining and assembly of all the parts to achieve and maintain proper lifter preload. Every machining operation we apply to the valves, valve job, and cylinder head deck has an impact on this measurement. While these operations are mandatory in a build, it's upon us to eliminate as many variables as possible. Keeping the stem tip heights close to equal is an important step.

Stem tip height is measured from the machined spring pocket on the top of the head to the tip of the valve. Special tools have been developed for this, but a good machinist's rule and a straight edge will also do the job. Insert a valve in each guide on a head and write the measurement down. You are trying to get them as close to even as you can. Sometimes it is possible to shuffle the valves from one cylinder to another and benefit from the machining tolerance variances to get them closer. While a racer will work toward perfection, a street build will generally be okay if you can get them all within .010–.015 inch of each other. If the variances are too

great, you may need to grind the tips of some valves to equalize them. Most valve grinding equipment has a fixture for doing this. You do not want to remove a lot; there are definite limits to how much can be removed, depending on the valve manufacturing technique. Once this process has been completed, mark each valve for location. They should be assembled in that particular chamber to keep the heights you've established.

Valve Stem Seals

The purpose of a valve stem seal is to control the amount of oil that travels down the valve guide. A small amount of oil is okay for guide lubrication, but too much will make for a surprisingly large amount of oil consumption, smoky exhaust, and mixture contamination.

Valve stem seals can take several forms. The original seals are referred

The valve seals you found during disassembly probably were similar to this umbrella-style rubber compound version. They are inexpensive, easy to install, fairly effective, and last a reasonable amount of time when new. Often with age they get hard and brittle, allowing too much oil down the valve stem and guide. In extreme cases, the debris from crumbling stem seals of this type can be found plugging the oil drain back holes in the cylinder head and the oil pump pickup screen.

to as umbrella style. They are simple inverted cups made from a rubber compound that are slid over the valve stem after the valves are installed. They are inexpensive, easy to install, effective, and last for a pretty long time when new. Umbrella seals stay snugged up on the valve stem and move up and down with the valve. As they age, they have a reputation for becoming hard and brittle and breaking off into pieces that get caught up in the oil pump. They are also large in diameter, making it difficult to run them with double springs. Because of this, a lot of folks go to an alternate design for performance builds.

Teflon seals were very popular with performance builders for many years. They are small in cross section, providing the spring clearance desired, and they are very wear resistant. They require machining work on the upper portion of the valve guide since they clamp around that diameter. The valve continuously slides through this seal design, which stays in a fixed location. These seals

The best seals to use are the Viton-type seals using a metal cylinder with an elastomeric insert formed in place. They are similar to the older Teflon design, but with an added degree of flexibility. While they do require special machining ahead of time, their effectiveness and durability is worth the effort.

have largely fallen out of favor over the past few years. They tend to be unforgiving in regard to guide wear, breaking into multiple pieces as a result of accumulated flexing in service. While the material is resistant to normal wear, any debris will scratch the sealing surface and reduce the seal's effectiveness.

Current best practice has been a move to a valve seal using a metal cylinder with an elastomeric (often viton) insert formed in place. These seals retain the small cross section of the older Teflon design, but add a degree of flexibility. These are both more durable and more effective than the other choices, but they still normally require guide machining prior to assembly. Plan ahead if you need or want these. Many FE heads may be able to accommodate viton seals with no added work, but it's far easier to check this now, rather than later when things are being assembled.

Cylinder Head Assembly

We are finally ready to assemble the reworked heads. The first thing is a very thorough cleaning; you need everything to be spotlessly clean and free of machining grit and debris. Using a bit of assembly lubricant or oil, slide the valves into their respective valve guides. They should go into the guides smoothly with no binding and slide smoothly up and down with minimal resistance.

If you have had the work done at a shop, you can spot check their work using a Sharpie marker or machinist dye. Smear some on each valve and seat and lightly spin the valve while applying closing pressure with your hand. You should see a clean and well-defined contact pattern on both valve and seat. (If the heads come

Here we are measuring the clearance between the bottom of the retainer and the top of the valve stem seal. Compare this to the cam's valve lift specification to make certain we do not have a conflict.

to you fully assembled, you can still check by filling each combustion chamber with solvent and watching for leakage past the valves.)

With the valves in place, install shims as required, locators, and then the valve stem seals. A dab of grease helps hold the loose parts together as you assemble them. The Teflon or viton seals need to be pressed into

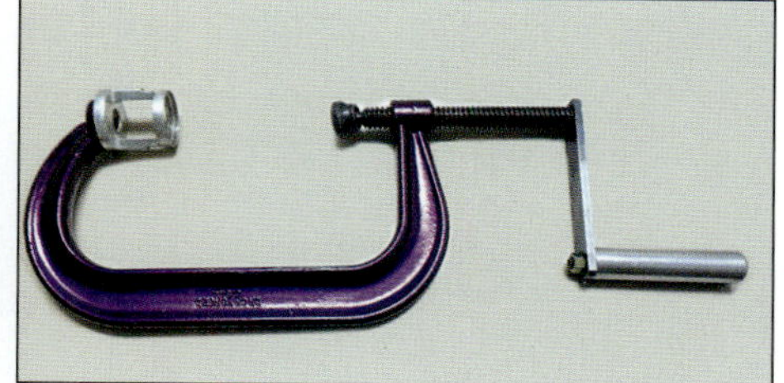

While there are many varieties of valve spring compressors on the market ranging from simple to complex, simple works in this case. The modified C-clamp-style of compressor is inexpensive and is more than adequate in almost all applications.

position around the guide using a small piece of tubing and finger pressure, or a light tap with a plastic hammer. You want the locators and shims on first because sometimes the seals are too large in diameter for the locators to slip over, and seals easily get damaged when you remove them. You should also check for retainer-to-seal clearance at maximum valve lift. You learn that lesson once.

To install the springs, retainers, and keepers, you need a spring compressor. Several types are available ranging from the popular "over center" levers, to a modified C-clamp design, to pneumatic cylinders for production shop use. My personal preference has become the simple C-clamp configuration. It is simple,

Installing the spring assembly is pretty straightforward. You simply place the retainer on top of the spring, position the C-clamp compressor over it, give a few turns until the spring compresses enough to install the keepers, then slowly release the pressure by unscrewing the compressor. The challenge is in holding all the parts in place and proper alignment. Be sure to check each assembly after releasing all compressor pressure to make sure everything is square and snug.

After you have installed all 16 valves and springs and double checked to make sure everything is secure, the heads are ready to be installed on the short-block assembly.

Before installing the cylinder heads, we need to install the locating split dowels. These help locate the head and head gasket in proper alignment as you install the head bolts and basically keep the heads from moving around. Install the split side into the block holes, lightly tap into place, and if any irregular surfaces or burrs are present, use a file to smooth things out to reduce the risk of damaging the head gasket.

inexpensive, and works very well on most projects.

Snap the retainer on top of the chosen spring, hold it in place over the valve, and mount the compressor around the head, keeping the retainer/spring assembly together. A few turns of the crank handle allow you to squeeze the spring down far enough to slip the keeper pair into position before you carefully release the pressure. It takes a little practice to hold all the parts together. Make certain that the keepers are in position and move on to the next valve.

With the heads completed, we can set them to the side and get ready to begin final engine assembly. We are getting pretty far along now.

Mounting the Cylinder Heads

To install the heads we first tap the locating split dowels into place. These are inexpensive and easily installed by angling the split side into the receiving hole first and lightly tapping the opposing side down until they are started. Once started, the dowels will tap into position with minimal effort. We sometimes find that we need to detail the protruding portion with a small file to get them free of burrs and defor-

mation; this makes subsequent gasket and head installation far easier.

For stock and modest builds we use the Fel-Pro standard replacement gasket sets. These come with a traditional composite head gasket with a formed steel fire ring around each cylinder. For higher-performance builds we move to the Fel-Pro performance gaskets, which have a flattened round wire embedded into the formed steel fire ring, making for a higher clamp load around each cylinder. This design gives more resistance to gasket failure from combustion loads, but at the cost of added cylinder deformation and some load-induced imprint on aluminum heads. On race builds we may use a Cometic multi-layer-steel gasket, but they are quite expensive and require extremely smooth deck surfaces.

This is the time to check piston-to-valve clearance. Although rarely a problem with a street-oriented FE package, checking this is a straightforward, simple process. The peace of mind makes it worth doing on any build project.

Our method for checking is to use a blob of common modeling clay on the piston. Knead and soften a chunk of clay large enough to fill the valve pockets and push it into the piston so that it sticks. We use the number-1 piston out of habit, but any of them will do. Spray the top of the clay or the bottom of the valve with silicone

Now that the alignment dowels are properly in place, we are ready to install the head gaskets. Make sure the sealing surface is spotless and that you didn't drip or smear anything onto the gasket mating surface that could put the proper seal in jeopardy. Make note of the large water openings at the front and rear ends of the head mounting surface.

so that the clay will cleanly release from the valves after testing.

Set the cylinder head on the deck being checked and attach it with a couple of head bolts. Just snug them up lightly, there is no need for a head gasket or torque wrench for this process. Now mount a rocker arm assembly and use a couple of lifters and pushrods set to eliminate all slack. Rotate the crankshaft a couple of times and the valves will compress the modeling clay.

Remove the rockers and the head. The squished clay will clearly reflect both the vertical clearance and the radial position of the valves. Take a handicraft knife and lightly cut a triangular slice from the clay at a few spots to see and measure the clay's thickness. Add the compressed head gasket thickness (a dimension supplied by the manufacturer) to arrive at the piston-to-valve clearance. As an example: If the clay is .100-inch thick and you have a Fel-Pro performance gasket at .041 inch compressed, you will have .141-inch piston-to-valve clearance. Some race folks will run tighter, but anything beyond .100 inch is a perfectly safe target for a street/strip package.

Changing this clearance is done by flycutting (deepening) the valve pockets in the pistons, or by altering cam timing. On a serious race engine where piston-to-valve clearance problems are common, we would have checked this far earlier in the build process, with only a single piston assembled into the block. But problems here are highly unlikely on a street FE and we are just checking for the sake of confidence.

We can now move ahead to permanently mounting the heads onto the block. Place the head gaskets over the dowels, paying careful attention to the location of the cooling passages in the gaskets. Each gasket is stamped with the word "front." The opening into the water jacket on the block must be blocked off at the front of each bank, and open in the back. This often means that the head gaskets will not appear symmetrical; one side's metal crimping will be facing "up" versus the other appearing to be "down." If you install the head gaskets backward, the engine will rapidly overheat. In addition to the stamped orientation note, each gasket will have a corner that protrudes out from the assembled head. If that corner of the gasket is not visible in the front of the engine, the head gasket is installed incorrectly.

Head Fasteners

You have a choice between using head bolts or head studs. Head studs are the preferred method for race engines, and are decidedly better for that purpose. But most street builds will be served perfectly fine with traditional head bolts. The build we detail here uses head bolts. (Plus, removing heads mounted with studs in any intermediate "shock tower" application will require significant engine raising or complete removal.)

The original factory head bolts are 1/2-inch diameter, and in 40 years I have never seen one fail. They do get pretty nasty looking over time, though, and replacement

When installing FE head gaskets it is critical that the water openings are open at the rear and blocked off at the front of the block. The gaskets are clearly marked "front" but it is easy to miss since it looks wrong to have the steel embossments and paint stripes facing up on one bank of the engine and down on the other. Front really does mean front. The front lower end of the gasket has a square corner protruding from the face of the head when properly installed. Get this wrong and the engine will quickly overheat.

Standard head bolts are 1/2 inch in diameter and almost never fail. Still, for more accurate torque readings, it's best to replace 50-year-old bolts. ARP makes nice replacements with the added benefit of smaller size head hex, which allows for more clearance, a plus for aftermarket aluminum heads. Whenever using aftermarket head bolts, refer to the manufacturer's torque specs and procedures; they are often different from factory. ARP also provides a special lubricant that needs to be used during installation and the torquing sequence.

Here you can see that telltale corner of the properly installed head gasket. If you do not see this corner sticking out there is a good chance that you have a gasket installed backward. Check it now before going further. If necessary, the head gaskets can be reused at this point since they have not seen any heat or pressure.

parts are readily available at a modest cost. You can also step up to a performance bolt set from ARP, which brings a couple notable advantages. They are stronger than standard fasteners and they have a reduced-size hex head. That smaller socket is a real advantage when installing aluminum heads, where space is at a premium.

If you are using head stud, screw them into the block. The head bolt holes on an FE are blind; no sealer is required. But you should use assembly lube and run them down lightly until they bottom. They do not need to be tight.

Next take a cylinder head and carefully lower it into position over the dowels (and studs, if used). Keep the head square to the block deck surface to engage the dowels into the recess in the head. Once started you might find that a light tap with a plastic mallet is required to get the head fully seated against the gasket and block. The aforementioned

light detailing of the dowels with a file usually eliminates this need. I generally keep a single head bolt at the ready when installing heads, just screw it in by hand after the head is in place so you can confidently move to the other side without worry.

With both heads in position we will install the head bolts. The head bolts need to be clean, the threads in perfect condition, and lubricated to deliver proper torque values upon tightening. Factory bolts did not use a washer, while ARP bolts come with dedicated washers. The washers for ARP bolts have a chamfer on the inside diameter that needs to face up, against the bottom of the bolt head to clear the radius ARP machines in that area. ARP studs also use washers, but those are flat on both sides and not location specific.

The recommended factory torque value for head bolts ranges from 95 to 110 ft-lbs, depending on the application and source of the data. We

usually go to 100 ft-lbs as a "normal" value, using ARP lube on threads and the underhead of the fastener. The recommended tightening pattern is essentially a spiral, starting from one of the center head bolts and working outward from there. We tighten the head bolts in a few steps. Tighten all the fasteners in order each time, moving up to the next tightness step after getting them all done. Our first is just to snug them up, maybe 20 ft-lbs of torque. Second is to get to around 65 ft-lbs, and then we go to the final torque value. You want to reach final torque with a single smooth sweep of the wrench, not with a short jerky motion. While not absolutely necessary, we have found it beneficial to come back to the engine after a couple of hours and loosen and retorque the head fasteners. This addresses the initial compression of the head gasket and elongation of the bolts, and seems to improve gasket life under duress.

VALVETRAIN ASSEMBLY

With the long block completed, we move ahead to lifter and valvetrain work. Roller lifters simply get a coating of oil and are slipped into their respective bores. The link bars for roller lifters traditionally go toward the center of the block. Although not normally an issue on factory blocks, be certain that the bars have clearance below them at all lift points. I have seen some that just hit the block in the valley area, causing a tapping noise that was nearly impossible to isolate.

Flat tappet lifters get a coating of cam break-in lubricant on their faces and oil around the outside diameters. It is critically important that flat tappet lifters slide easily into their bores and are able to freely rotate. A lifter that binds in the bore will not rotate and will likely cause an eventual cam failure.

On flat tappet lifter combinations we will install the factory valley tray after the cylinder heads are installed. The tray is a formed sheet-metal piece that controls oil splash. It is a nice thing to have, but cannot be used with roller lifters because it interferes with the link bars. The tray snaps into position under a series of tabs on the head gaskets. You start it on one side of the engine by slipping it under the tabs and then snap it into place on the opposite side. Once the tray is snapped in, use a small rubber hammer and fold the tabs on the head gasket down to retain it.

Engines originally equipped with hydraulic lifter camshafts have non-adjustable cast-iron rocker arms, with a machined receiving cup at one end for the pushrod and a radiused pad on the other end for contact with the valve tip. Solid lifter applications have an adjustable rocker arm, similar in design but including an adjuster screw at the pushrod end. The adjusting screw will have a ball end that requires the aforementioned cup-type pushrod. The threads on the adjuster screw are an interference type, designed to be difficult to turn and to remain in position once adjusted. After multiple adjustments the factory screw

If you are using roller lifters, make sure the link or attaching bars point to the middle of the lifter valley as shown in this picture. You will want to double-check these as we get further along in assembly to make sure there is no interference with the tie bars and the lifter valley.

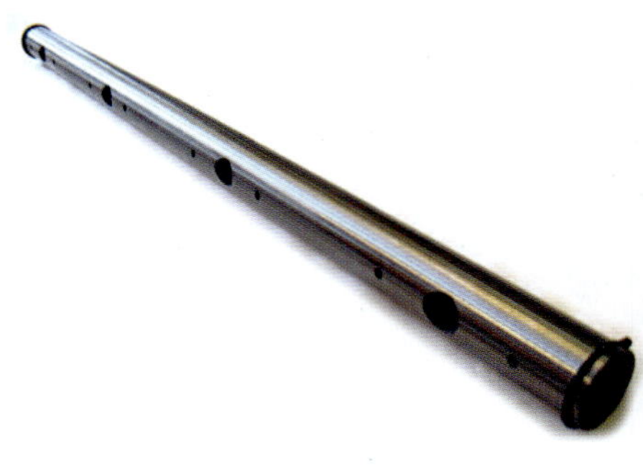

If you have a lifter valley tray, use it. As you can probably guess, the tray fits only OEM-style flat tappet lifter configurations; the tie bars on roller lifters cause interference.

Most rocker arm shafts have a single row of oil feed holes at each rocker position. These holes need to be assembled facing down toward the head to provide lubricant at the loaded area of the rocker arm.

The rocker systems in a Ford FE comprise a single shaft for each side with four mounting stands attached to the cylinder head with four 3/8-16 fasteners. On stock rocker systems the rockers are separated by springs.

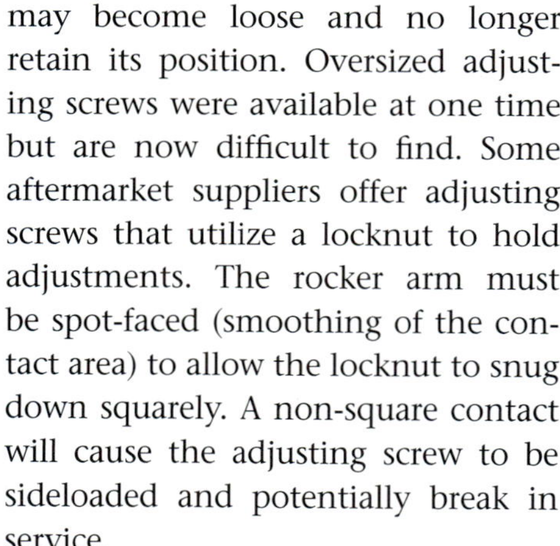

The end exhaust rockers on an FE are unsupported; they are cantilevered off the end of the shaft. Higher spring rates and more aggressive cam profiles make the shafts prone to breakage as a result. Billet aluminum end stand kits provide the reinforcement needed for greater durability and are good insurance.

may become loose and no longer retain its position. Oversized adjusting screws were available at one time but are now difficult to find. Some aftermarket suppliers offer adjusting screws that utilize a locknut to hold adjustments. The rocker arm must be spot-faced (smoothing of the contact area) to allow the locknut to snug down squarely. A non-square contact will cause the adjusting screw to be sideloaded and potentially break in service.

Both types of factory rocker arms ride directly on the shafts with no bushings or bearings. If there are signs of significant wear on the valve tip end or on the inside diameter that rides against the shaft, the arm should be replaced. It was at one time common to see bushings

installed to rebuild the inside diameter, but that process reduces the cross section of the arm and compromises arm strength.

Rocker shafts for the FE are made from steel, usually with a chromed wear surface for durability. Any scoring or damage to the shaft mandates replacement. Stock replacement shafts are inexpensive and work fine for most street builds. Heavy-duty shafts are available from several suppliers and are a good upgrade for more serious performance applications.

The majority of original FE rocker stands are cast aluminum, with four identical stands used to attach the rocker assembly to the heads. A close inspection will show that the upper mounting hole of each stand is sized to accommodate the attaching fas-

tener, while the lower portion of that same opening is larger in diameter or oblong shaped to promote oil flow from the head into the rocker shaft. On factory systems this works in concert with a necked-down mounting bolt to deliver oil from the cylinder head to the rocker shaft. The stands used in 427 Ford engines were steel for additional clamping force and strength, but those are difficult to find these days. Aftermarket stands are available that wrap around the end of the rocker shaft to provide additional support. Rocker shaft breakage is fairly common in high-performance usage, and the "end stand supports" are an excellent upgrade for added durability.

Pushrod Length Measurement

Pushrod length is critical on any FE build. It is even more so when running the factory nonadjustable rocker arm system. It is easiest to measure for pushrod length now, before installing the intake manifold. While the normal replacement pushrods may be perfectly adequate for stock or near stock builds, the variables of machining and camshafts make it a far better policy to measure and order exactly what you need.

To do this, once again temporarily mount a rocker arm assembly to a cylinder head. Rotate the engine until the chosen lifter is on the base circle (the lowest point) of its cam lobe.

You will need to acquire or fabricate a measuring pushrod. This is a tool that has a threaded center section allowing you to extend its length by rotating it. These are available from numerous sources or can be made with an old pushrod cut in two and a small piece of threaded rod.

With adjustable rockers you need to set the adjuster to its desired operating position. On most aftermarket rockers this will be with two or three threads showing below the rocker arm body. On factory adjustable rockers perhaps four or five threads will show. The idea is to prevent the upper pushrod cup from contacting the rocker arm at any point throughout its travel.

Measuring push rod length is critical in all FE builds, but especially if you are running the nonadjustable rocker arm system. You can make your own adjustable push rod for measurement purposes, or purchase one from many aftermarket sources for under $20. With the rocker arm lightly seated against the valve tip, unscrew the length measuring pushrod until it lightly contacts both the rocker arm end and the lifter end. Note that the FE uses a 3/8-diameter ball and cup at the rocker arm end.

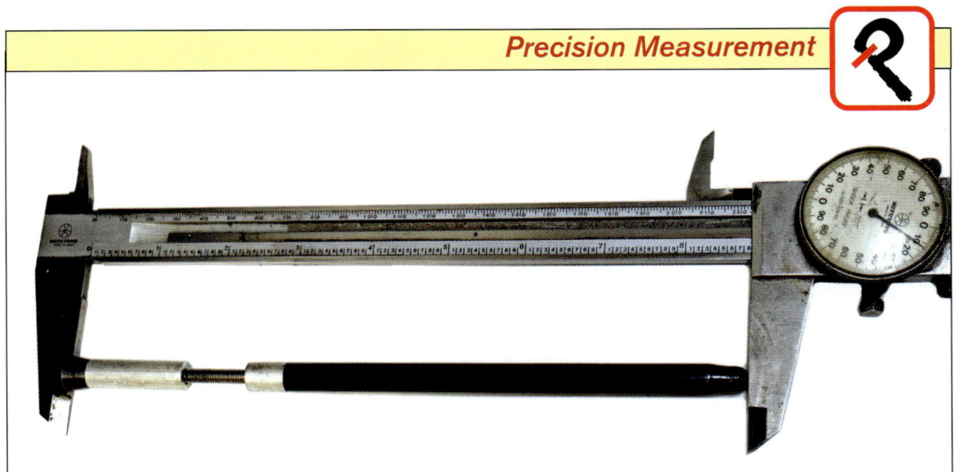

Once you have determined the proper length using your adjustable pushrod, measure using a 12-inch long dial caliper and record the length. Remember to account for the ball diameter if using cup-style pushrods, and let your pushrod supplier know you are measuring from the bottom of the cup. For hydraulic applications, add .050 inch to the length to account for preload. Solid lifter applications will use the exact measured length.

With the rocker arm lightly seated against the valve tip, unscrew the length measuring pushrod until it lightly contacts both the rocker arm end and the lifter end. Loosen the rocker assembly and set it aside. Gently remove the measuring pushrod tool without disturbing its adjustment.

Use a 12-inch-long dial caliper to measure the length of the pushrod tool. On hydraulic lifters you will want to have approximately .050 inch of preload for final adjustment; add that .050 inch to the

Now would be a good time to check for rocker arm tip alignment as well. Set the rocker arm assembly back onto the head (without any pushrods) and lightly snug it down. Look at each rocker where it would contact the valve. It's often easy to add a valve spring shim to shift the rocker over a little bit to better center the rocker.

Factory FE engines use a series of springs to separate the rockers from each other. On high-performance builds the springs may be replaced with solid spacers for more positive location.

specified pushrod length so that the adjustment screw remains in the same place as when measuring was done. For solid lifter applications the exact measured length will be sufficient.

Because of design variations, such as cups or oil holes, there are several ways to specify pushrod length: theoretical, actual overall length, or bottom of cup. More important than the method you choose to employ is the need to clearly state your measurement technique to the company supplying the pushrods. All the major pushrod suppliers can take it from there as long as you can clearly state the dimensions you need and the method you used to reach that number.

Make certain to let your supplier know that you are working on an FE. The FE uses a 3/8-inch-diameter ball on the rocker adjuster, while other

engines use a smaller, 5/16 inch.

This is also important if you decide to use factory nonadjustable

rockers; that 3/8-inch ball size is not very common. Hydraulic roller combinations can be used with non-adjustable factory rockers as long as proper length and style pushrods are ordered. The hydraulic roller lifters are designed for a 5/16-inch ball diameter at the lifter, so pushrods will have two different ball diameters.

As a general reference, common lengths for FE pushrods will be

Check the valve-train geometry for proper rocker–to–valve-stem-tip contact area. Ideally you want the rocker arm to sweep over the center one-third of the valve tip, and not move toward either edge. This is more easily seen when using roller rockers. Unlike in many other engines, pushrod length will have a minimal impact but adding a shim under the rocker mounting pedestals can help.

8.350 inches to 8.500 inches when running a hydraulic roller cam. A flat tappet cam will usually require pushrods be quite a bit longer, often 9.000 to 9.350 inches. Solid roller parts can vary considerably depending on the lobe base circle and the rocker system selected. My personal preference is for a minimum 3/8-inch diameter and .080 wall thickness for the hydraulic roller or flat tappet pushrods, and 3/8-inch diameter and .135- to .165-inch wall thickness for solid rollers. A thin-wall pushrod will deflect in service and cost you valve lift and power.

This is the appropriate time to check for rocker arm–to–valve tip alignment and for valvetrain geometry if you stray away from a basic factory rebuild. On stock FE rocker systems, the rocker arms are separated on the shaft by springs. On aftermarket systems, we use solid aluminum spacers. Rocker arm position relative to the valve tip on both systems can be adjusted to an extent by running a shim between the rocker arm and the adjacent mounting stand.

Larger corrections may require machining to the sides of mounting stands or solid spacers, when used. Really significant changes required to accommodate for radical ports and such fall into the realm of race car stuff, and outside the context of this book.

Valvetrain geometry can also be checked now, assuming you have appropriate lifters and pushrods on hand. The basic concept is that you want the rocker to sweep across the valve tip in a very narrow area, and not to go over the edge. We will first paint the tip of both valves on a chosen cylinder with a Sharpie or machinist's dye. Then we assemble a complete rocker sys-

The FE engine design is unique in that the pushrods are routed through holes in the intake rather than in the cylinder head. Unlike other engines where you can install and adjust the entire valvetrain before installing the intake, the intake must be installed first on the FE. The only real challenge this presents other than workflow is your ability to make sure the pushrods are properly seated in the lifters before final adjustments. You also must make certain that the pushrods are not bound up or hitting on the manifold itself during rocker travel. A flashlight comes in handy for this step.

tem for the cylinder with pushrods and all in a reasonable approximation of final adjustment. Rotate the engine through two full revolutions by hand to simulate operation through a complete cycle. Remove the rocker assembly and you will see the contact pattern. It should be reasonably centered on the valve tip and perhaps .080 to .100 inch wide.

Unlike with many other engines, pushrod length has a minimal effect on rocker geometry. If you need to change geometry for some reason (unusual on a mild build) you must either shim the rocker stands up or mill them down. If changes are needed a close observation from the side of the rocker as it travels its arc will show which way you need to go.

A few competing theories and philosophies about the "correct" geometry for a race application are out there, but as long as you stay on the center third of the valve tip you'll be safe on a common street build.

Since the FE has pushrods that travel through the intake manifold, we now need to install the intake, so things are going to get a bit out of sequence here. Skip ahead to the intake manifold chapter and then come back here to resume the valvetrain work once the manifold is bolted down.

Assembling the Valvetrain

Pushrods are installed by putting them through their respective holes in the intake manifold after putting

The factory valvetrain on the FE uses a drain tin for oil control, which helps guide the oil into the valley. The factory tin does not play well with many aftermarket rocker systems or intake manifolds because of interference. It's not the end of the world if you end up not using it, but if you have one that fits, you should use it.

Direct replacement high-performance roller rocker arms are most often aluminum extrusions. Designs that run the aluminum directly against the shaft are common and less expensive, while others incorporate a bronze bushing or a needle bearing fulcrum, such as those on the high-quality T&D system shown. Most aftermarket rockers use a locknut and threaded adjuster for valve adjustment, and have a ball end similar to the factory adjustable arms. Several manufacturers sell complete FE rocker systems that include the rockers themselves, along with dedicated mounting stands, spacers, and shafts.

a dab of lube on the tip. A flashlight can come in handy here, allowing you to verify that they are properly into the cup in the lifter center. It is surprisingly easy to get them off location. If using factory nonadjustable rockers and a hydraulic roller cam and lifter combination, the larger 3/8-inch-diameter ball should be at the top. Don't forget to put a dab of oil on the tops of each pushrod before moving on to the next step.

The oil feed passage for the valvetrain intersects one of the center rocker mounting holes of the cylinder head. We use oil feed restrictors on a lot of our engine builds. Restriction is pretty much mandatory with aftermarket rockers having roller bearings on the shaft, but optional with bushing-style and OEM rockers. They can easily be installed later if you decide to add them. They often add roughly 5 psi of idle oil pressure and reduce flooding of oil in the valve covers if that is a problem.

The restrictors are inserted into the oil feed passage in the heads before attaching the rocker arms. On most factory iron heads, a 70 jet from a Holley carburetor works well; just slip it into the passage. There is no need to worry about them staying in place; they get trapped in between the head bolt and the rocker fastener and cannot come out. On aluminum heads we have to fabricate a restrictor to slip into place.

Factory valvetrain systems use a drain tin under the rockers that helps guide oil to drain through the holes in the intake and down into the valley. I use them when I can. Two popular tin designs with either long or short fingers direct oil flow. The long fingered style works well but will not fit on many intake manifolds. The short finger style fits almost all intakes, but neither one will work with many aftermarket rocker systems. If you can't get one to fit it's certainly not a huge deal, but they definitely help with oil return flow. The drain tins are simply set in place and get captured by the rocker stands when fastened down. Many aftermarket roller rocker systems do not allow you to utilize the drain tins due to interference.

The rocker assemblies may be mounted to the heads with either factory bolts or with aftermarket studs. I prefer studs on aluminum heads because they are easier on the threads. Similar to the bolts, thick and strong flat washers should be used with studs, although most of the available studs do not have the necked-down shank section. All fasteners get lubed before assembly.

Whatever fastener you choose, the assembly method is similar. Carefully lower the rocker assembly into position and move the pushrods into the receiving cup or adjuster screw on each rocker. This can be a challenge, but you need to be positive that none of them are popped loose from their proper location or you risk damaging parts, or at the very least redoing the installation. If you are using studs, lower and align the rocker stands over the studs while trying to keep everything level. If using bolts it helps to drop the bolts through the stands and lower the assembly toward the head. A careful touch and an extra hand is useful here as the stands and rockers always want to move around.

With the rocker assembly roughly in place, start tightening things down, going back and forth from fastener to fastener a little at a time, only a quarter or half turn. The idea is to keep the rocker shaft and stands parallel to the head as you go. You do not want to run one end down first or to put any sort of bending load on the rocker shafts. Once you have the assembly completely snugged down against the head, take a close look to ensure that none of the pushrods has escaped from the desired position. Torque the fasteners to specification. On iron heads with bolts I use the factory specs; on aluminum heads with studs I tend to lighten up a little and run no more than 40 ft-lbs.

If you are using nonadjustable rockers, you are now done with this part. If you are using adjustable rockers, it's time to set the adjustment.

If your chosen system allows rocker adjustment, you can go about it a few ways. I prefer to go one at a time and mark each rocker as I cycle through. Rotate the engine to bring the chosen rocker to the lowest valve lift position, the base circle of the cam. Loosen the adjusting screw and locknut (if equipped).

On a hydraulic lifter application, I now lightly take up the clearance until I just feel the pushrod get snug as I wiggle or twist it. Then I go down *one full turn*. This is not a traditional small-block where you have adjustment at the fulcrum; on an FE engine you adjust lifter preload directly at the pushrod. You want about .050 of lifter preload, and a full turn gets you reasonably close. If you try a quarter turn you will get only .010 of preload and will lose much of that as the engine warms up, which is a recipe for ticking lifters.

The factory mounting bolts for the rocker arm shafts are of two different lengths, with a single longer fastener having a necked-down center section for oil fed from the passage on each head. They use a very thick and strong flat washer to provide support and prevent crushing the aluminum stand.

On solid lifter applications I go to the specified lash minus .008 inch on aluminum heads or minus .004 inch on iron heads as an initial setting. If your cam specifies .016-inch lash, set it to .008 inch for an aluminum head engine. These values are based on experience. Everything grows as it heats up so it is best to reset once up to temperature. Your actual numbers may differ, but this gets you darn close for start up.

Continue setting lash or preload going one rocker at a time, marking each one as you go. This also gives you a chance to look for potential problems. An adjuster that ends up at a significantly different height than others, popping or snapping noises during rotation, or binding and tight spots are all indications of issues that require attention. Even if finding and fixing a problem involves significant disassembly, its way better to catch it now rather than after something fails dramatically in a running engine.

INTAKE MANIFOLD SELECTION

A huge variety of intake manifold options for an FE engine build exist; most of them are out of production now and would be swap meet and private purchase items. The choices are far more limited for parts that can be acquired new.

The FE enthusiast community is divided into three general segments, and the right intake for you depends on your place among them. At one end of the spectrum are folks looking for originality, whether driven by budget or the need for a restoration-mandated part. At the opposite end are folks fixated on maximum performance at any cost. And through the middle are most enthusiasts, who weigh the variables of cost, cosmetics, and performance against a personal budget.

The restoration folks are going to procure the right part, so that decision is made for them. The racer types are going to migrate toward the Edelbrock Victor or the tunnel wedge dual quad intakes, whether new or reproduction.

The book *The Great FE Intake Comparison* by Jay Brown is the definitive document regarding comparative performance among virtually every intake manifold ever produced for the Ford FE engine. Exhaustive dyno test results for multiple FE engine combinations are included, along with descriptive and dimensional data. If you are willing to seek out used and vintage parts, this book is a great resource and well worth referencing. You will discover that some vintage single plane manifolds, such as the Edelbrock Streetmaster, perform amazingly well.

For the sake of brevity, I limit this portion of the discussion to currently available manifolds targeting a street-style FE build. The street-oriented single 4-barrel manifolds currently available new are the Edelbrock Performer, the Edelbrock Performer RPM, and the various Blue Thunder iterations. All are dual plane designs.

For a numbers-matching or period-correct restoration, your choices for intake manifolds are obvious. While factory 4V dual plane intakes are respectable performers, it is difficult to recommend any cast-iron intake unless factory correct is your reality. Cast-iron intakes easily weigh 40 pounds more than their aluminum counterparts.

The Edelbrock Streetmaster is an intake that has been out of production for many years. A single plane design with small runners, it performs extremely well on smaller displacement street engines.

The Performer RPM could be considered the universal manifold for street-performance applications. It works very well in almost all situations, flowing enough for good performance, priced affordably, and readily available, a consideration in the FE world. Note that it will cause fitment issues in shaker hood scoop applications, and has no provision for the oil tube found in early FE applications, requiring different valve covers to make it work.

Although more expensive, beautifully detailed Blue Thunder intakes are more accommodating of all the OEM nuances, such as linkage mounts, oil fill tubes, vacuum ports, and carburetor positioning. The performance potential is equal to or slightly better than the Performer RPM manifold's.

Current (or recently) available dual 4-barrel intakes are the Blue Thunder products and the Edelbrock Performer RPM, which are dual planes, or the single plane tunnel wedge intakes from Dove and BBM.

The Performer is a fairly low-cost option with very small ports and rather simplistic machining. It is best considered a "want a 4-barrel" type of intake. Performance is on par with common factory cast-iron manifolds, with the primary advantage being that it is 40 pounds lighter and a lot easier to wrestle around.

The Performer RPM is perhaps the best intake available on a power per dollar basis. The ports are generous enough to handle a fair amount of power with minimal work, the cost is modest, and availability is excellent. If the application allows, you cannot go wrong in choosing this manifold for any performance-oriented FE project. There are a couple limitations though. If you are running an early-style engine and want to retain the smooth valve covers, the manifold-mounted oil fill tube, and the rear of the intake breather, the RPM simply will not work. Also, it will pose fitment challenges on Mustangs equipped with a shaker-style hood scoop, as the carburetor mounting pad is higher and farther back compared to the original manifold.

This brings us to the Blue Thunder offerings. These intakes are available in several styles and port layouts, but all are based on a common casting. The Blue Thunder single 4-barrel intake is the only one that combines current (albeit limited) availability with accommodation for all the OEM possibilities for oil fill tube, throttle linkage mounts, vacuum ports, and proper carburetor

Also available from Blue Thunder are dual quad, dual plane manifolds. The Blue Thunder accommodates Holley carbs. Nothing looks better than multiple carbs on a big-block FE engine with the iconic finned oval air filter. The performance benefits on the street are real, but modest.

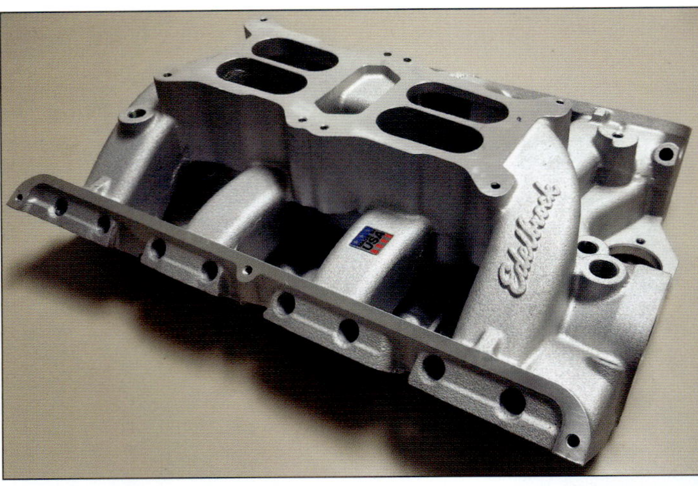

The newest FE intake on the market is a dual quad, dual plane offering from Edelbrock. It appears to be a very good item from a performance perspective, with a modern design and an air gap below the runners to keep induction temperatures low. The company decided to make it accommodate only the Edelbrock carburetor, so direct comparison to the Holley-based intakes is difficult.

position. Performance potential is at least equal to that of the Performer RPM, if not better. Casting and machining quality are among the best in the performance industry. The only downside to the Blue Thunder manifold is a higher cost than the comparable Edelbrock offering.

Although outside the main focus of this book, it's worth mentioning that two dual plane, dual quad manifolds are also available new for the FE family. One is from Blue Thunder, the other from Edelbrock. The Blue Thunder parts share the same excellence in design and build quality as their single 4-barrel offering. The Blue Thunder 2 x 4 is an updated approximation of the Ford OEM dual quad intake from the mid/late 1960s and accommodates the iconic dual Holley carbs and oval air filter that have become so strongly associated with the FE engine. The Edelbrock dual quad is a completely new air gap design, and

will fit only the Edelbrock/Carter–style carburetors.

Fitting the Intake Manifold

Intake manifold fit and sealing is a huge focal point on FE engine builds. The FE intake is the only intake on a modern V-8 that has the intake gasket completely enclosed under the valve covers, which is an oil-soaked environment. Any flaw in gasket seal will cause the engine to inhale large amounts of oil, fouling plugs and putting out a cloud of blue smoke. My process for installing the FE intake is time consuming, but near mandatory to ensure success. Forcing it to fit by tightening it down with bolts will result in leaks and broken and cracked bolt bosses.

The very first thing to do is to make certain that the gasket faces on both the heads and the manifold are perfectly clean and flat. We are going to assume that the heads

have already been addressed, but even a brand-new intake needs to be checked and possibly corrected. We have seen many out-of-the-box manifolds that did not have straight sealing surfaces. They can be quickly checked with a good-quality straight edge and a feeler gauge. Lay the straight edge along the gasket flange surface and probe with a .003–.005 gauge; any place that allows the gauge to slip in will cause a sealing problem and needs to be addressed by machining the surfaces flat.

With everything clean and flat, carefully set the manifold in place using a pair of non-embossed intake gaskets, such as Mr. Gasket or Blue Thunder parts. Embossed gaskets are perfectly fine for final assembly, but not usable for this mock-up measurement. If you are sticking with a cast-iron intake manifold for originality or budget purposes, it is often good to get an extra set of hands to help lower what feels like the world's heaviest intake into place.

Intake Manifold Alignment Test Fit

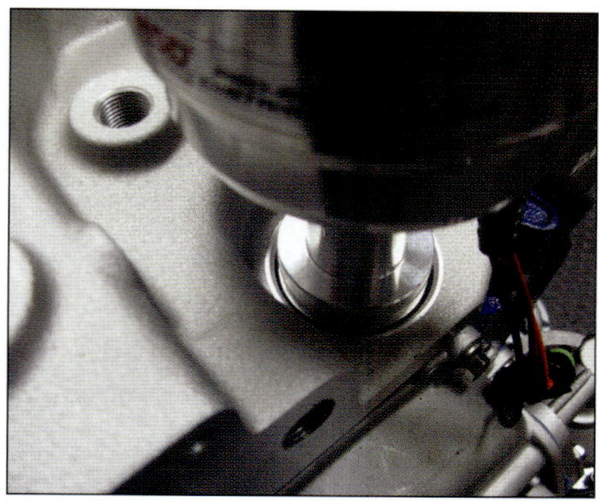

1 Use the distributor as an alignment tool; the manifold must be perfectly centered around the distributor, and squared up as flat and even as possible relative to the end seal surfaces of the block. We have numerous areas to measure and inspect, so try not to disturb the manifold once positioned. Before measuring, make certain the manifold is truly sitting on the manifold gasket surfaces. Some aftermarket manifolds have excess casting thickness below the ports and will stand proud on the head gasket edges in the valley instead of dropping properly into place. You may need to remove material from below the ports on the intake or bend the protruding head gasket tabs downward.

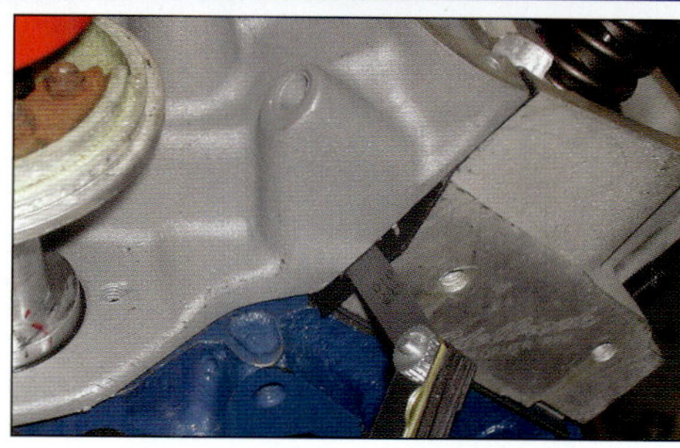

2 First we check for gasket face angle. Starting with a .010 feeler gauge, try to slip it in between the manifold and cylinder head in multiple places, upper and lower front left and right, and upper and lower rear left and right.

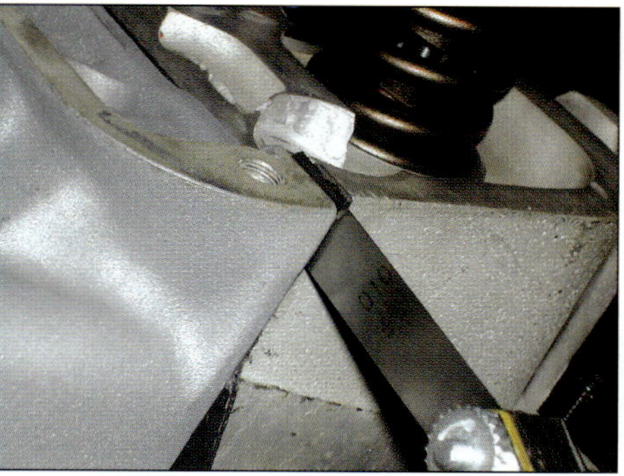

3 If the manifold is tight at one spot and slides in freely elsewhere, you are going to need to resurface the manifold to get the angles to match up, or it will eventually leak. I use a Sharpie marker to make notes on the manifold itself where the measurements are and reference the modifications needed. It is not at all unusual to need the angles altered because of tolerance stack up.

4 If the gasket flange angles are incorrect, this can be the result. This shows excessive crush on the lower portion of the gasket and minimal contact on the upper side. The crushed part split and allowed oil into the port under vacuum. Lots of blue smoke and poor running are the all-too-common results.

5 Assuming the angles are either correct, or have been corrected, we next look at the relative location of the intake mounting holes and the valve cover gasket surfaces. Using a small flashlight, you can sight down the mounting holes to see if the threads in the head are centered. If they are at the upper edge or above, the manifold is too wide and gasket flanges need to be milled down so that it will nestle down farther into the valley. If they are too low, a doubled up or extra thickness gasket set may be required.

6 Next is to check the alignment of the intake's valve cover gasket surface to those of the cylinder head. Usually these are misaligned when other fitment issues are found and fall into place once those are corrected. The common cork valve cover gaskets are thick and forgiving compared to other gaskets, so a little misalignment won't hurt as much. If everything else is good but these are significantly off, have the valve cover gasket flanges machined or possibly modify a cork gasket to compensate.

7 *Our final intake check is to verify pushrod clearance. Rarely a problem on mild builds or when using original parts, but something that can really mess things up as you stray into greater modification. On the FE engine the pushrods must pass through holes in the manifold. Test fit the intake and assemble the valvetrain into position. Rotate the engine through a couple revolutions and carefully look for any signs of contact or interference between the pushrods and the manifold. Any contact must be addressed now by shimming rockers or modification to the manifold, or you will be taking things back apart later.*

With practice and careful calculation, the required manifold machining can be sorted and done in only one or two steps. Take careful notes. I write the measurements and a little drawing on the intake itself and take it to the machinist. A well-fitted intake will drop right into place, fit square, and all the fasteners will start with your fingertips.

If you plan to port match your intake manifold, this is when you mark out the top, bottom, and sides of the ports on the head and transfer that data to the intake. Concentrate on the roof and sides of the port, and go in at least 2 inches or so. Keep away from the pushrod passages. If you do break through a pushrod passage, you can epoxy in a brass or aluminum sleeve as a repair.

Installing the Intake Manifold

1 *With all the work completed and the intake cleaned up we can move ahead to final installation. I use an extremely small amount of TA-31 silicone around all the water and port openings. Using a fingertip, I apply a layer as thin as a coat of paint; you can practically see through it. It's really being used as insurance against any possible leakage. Theoretically the manifold gaskets can go on dry if everything is perfect, but one "leaker" and the subsequent rework on an installed engine will steer you toward using the sealer.*

2 *Next, position the chosen intake manifold gaskets in place. My preferred gaskets for Edelbrock or factory small port heads that have stock port openings would be the Fel-Pro 1247-S3, which has a metal layer bonded through its center. For the larger port openings found on low-riser and Cobra Jet heads, I use either the similar 1246-S3 or the lower-cost MS90145 passenger car replacement gasket, which works perfectly fine. If the ports have been opened, or if you are using a larger-port aftermarket head, use a Blue Thunder gasket, which can be trimmed to fit and is still very resistant to oil soak failure. The traditional Fel-Pro race gaskets without the metal core will fail in street service on FE applications and are to be generally avoided on builds with an expected long service life.*

3 *Cork end seals have become a personal preference item. Some folks still insist on using them, while others have gone to using a large bead of silicone. I am of the latter opinion and take advantage of the caulk tube employed by my favorite TA-31 Motorcraft silicone to run a bead across the front and rear end seal surfaces. It helps adhesion to wipe a small amount on the rails with a fingertip before squeezing out the bead. Run the bead up alongside the intake gaskets a little bit to seal up the corners where everything meets. Another thin, thin layer of silicone on top of the manifold gaskets and I am ready to set the manifold into position.*

4 *Lower the intake down as straight as possible to avoid messing up the silicone (or cork) end seals. Once in position you again use a distributor to center and locate the manifold on the block. Start all the bolts by hand, using a bit of lube on the threads. Make sure to use a flat washer under each bolt head, and a tiny dab of silicone under the washer to help prevent oil from migrating up onto the intake. Once everything appears straight and square, you can tighten the manifold fasteners.*

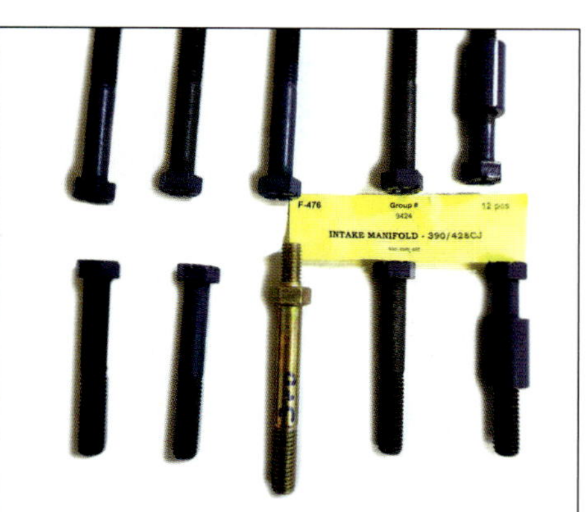

5 *Despite the torque specs, it is challenging to get a torque wrench on all the fasteners. Follow the torque sequence in the appendix. Torque the ones you can reach, and it's perfectly okay to simply get the others good and snug using a box wrench once you have a feel for how tight they need to be. Note the specialized fasteners on this package. The one with the stud is for the driver's side center, for mounting a factory shaker hood scoop. The ones at the rear on the manifold use a sleeve to make up the difference in length.*

6 *Remove the distributor and use a fingertip to carefully remove any silicone that squeezed into that area. You do not want anything to bind the distributor up later. Now is a great time to walk away for an hour or two and let the sealant cure before disturbing things.*

Once the intake is in place and silicone has dried, install any accessories your application may require, such as the oil filler tube or cover for the rear breather opening. Now you can return to installing the valvetrain and rocker assemblies.

CARBUROTORS

Most builds following this book are going to have a 4-barrel carburetor. Over the years of production, Ford installed both Autolite and Holley 4-barrel carburetors on the FE family of engines.

Odds are that only those folks who already own one, or those who are working on a restoration, are going to be running the Autolite carburetors these days. Autolite 4-barrels perform perfectly fine on a mild build and were the standard 4v carb on most nonperformance FE engines throughout the 1960s.

The factory-supplied Holley carbs on the 390 GT or the 428 CJ are excellent units with superior calibration from the factory. If you have one, by all means use it. The original Holley carb for these is a 4150 configuration. It has a unique front fuel bowl with an inlet fitting and filter screen on the front driver's side and a cross feed tube on the passenger's side, which provides fuel to the rear bowl.

Ford also offered dual quad and three 2-barrel options, but these are rare and expensive. We discuss them a bit, but really focus on the single 4-barrel.

For new carburetors, we are most likely looking at Holley, Edelbrock,

Builders seeking originality will opt for the factory-original Autolite carburetor in cases where a Holley was not equipped. While the Autolite 4100 models were popular as a decent performer on mild builds, the 4300 model was more difficult to get dialed in correctly because of new emissions concerns, and the early sizes of the 4300 were often too small to meet engine demands. Neither model Autolite was made in a size bigger than 600 cfm, an obvious restriction to performance.

Holley provided the excellent 735 cfm 4150-style Holley for 428 CJ applications, and they are excellent carbs. Keeping your numbers-matching 428 CJ with this carb is not a performance impediment at all. These Holleys are well tuned for the application and are a very solid and serviceable design. A similar 600 cfm Holley was installed on 390 GT engines.

The Holley 4160 series of carbs have remained the gold standard over the past 40 years. They are affordable, and for mild applications, close to what you need right out of the box. The secondary metering plate does limit tuning options, so if you are looking for performance tuning on the secondary side, either add a secondary metering block or select a 4150-style Holley in the first place. The 4160-style carbs are traditionally vacuum secondary affairs, while the 4150 style of Holley could be vacuum or manual secondary (double pumper) in design. Factory-style dual quads mandate the 4160 carbs because of space; a 4150 will not fit.

If originality is not a concern, the Edelbrock series of carbs, based off the old Carter design, is a valid choice. Edelbrock keeps updating them, and they are very tunable. They also come in a variety of sizes. (Photo Courtesy Edelbrock LLC)

Multiple carburetors add a certain wow factor on the street, but come with added cost and complexity. Ford used dual quads and tri-power setups from the factory, and with proper tuning, these systems performed and behaved very nicely on the street. Note the lack of space between the two 4160-style Holleys. This is the reason 4150-style Holleys will not fit; they are too long because of the secondary metering plates. The carbs shown here are a pair of 660 center squirters, vintage race car stuff.

or Quick Fuel products (the latter is similar to a Holley in terms of basic architecture). All three use the 4150 flange pattern and will physically bolt on to most popular factory or aftermarket intake manifolds. The Holley design has a combination of traditional cosmetics coupled with great power potential and a lot of tuning flexibility.

The Edelbrock carburetor traces its lineage to the Carter AFB and AVS carbs of the 1960s, and some folks like the design and functionality. Having worked at Holley for a decade, I admit to a personal preference for those. That noted, the Edelbrock does have a reputation for good street manners and easy tuning.

On a street-dedicated build, a carburetor with vacuum secondaries is the best choice for gas mileage and driving behavior. A 600 cfm version is adequate for a very mild near stock build, but a 750 is a better choice for performance usage, especially on larger engines. The addition of a metering block on the secondary side of lower-cost 4160-style Holley carbs is a very good upgrade, as it provides far greater tuning flexibility.

The 4150 series of Holley carbs comes with that metering block, and a wide variety of specialized high-performance iterations are available. The double pumper carbs use a mechanical secondary design with an additional accelerator pump, and are best suited for serious high-performance work in lighter cars having either a stick shift or a looser-than-stock torque convertor.

The carburetor is mounted to the intake with four studs, most often with some sort of spacer between the manifold and carburetor. A spacer almost always shows a benefit on these engines, and the factory used

Unique to the factory dual Holley setup is that the carbs are mounted backward compared to other applications. The primaries face the firewall, special linkage puts throttle operation on the passenger's side, and the vacuum secondary canisters are connected by a vacuum hose to ensure even operation between the two carbs.

Quick Fuel Technology makes quasi-reproductions of the OEM factory setup, which run with factory mountings just like the Holley versions. The linkage for both brands is progressive, which allows for better street drivability. While the Holley looks more traditional, the Quick Fuel offers greater tuning flexibility.

one on most original applications. While claims are made for the benefits of open, four-hole, and transitional designs, the only way to truly know what the engine wants is to try them on the car. Every combination is different.

Dual quads are a signature item for the high-performance FE engines of the 1960s. Although expensive, these systems can be made to perform and drive very nicely in a performance-oriented street application.

The factory dual quad intake manifolds are often found online or in swap meets. The most common ones feature low riser ports. The less common medium riser itera-tions are usually more desirable and more expensive. Blue Thunder also manufactures a very nice upgraded reproduction-style dual quad intake with either port configuration along with numerous vacuum openings to accommodate PCV systems and power accessories.

Carbs may be either originals that need a restoration, reproductions, specialized high-performance variations, or generic replacements. The carbs on factory-style installations are Holley 4160 designs with vacuum secondaries. Unique to the FE is that the carbs are mounted "backward," with the primary barrels facing toward the firewall and the linkage connection on the passenger's side of the engine. A vacuum hose connects the vacuum secondary diaphragm canisters to promote an even opening rate between the two carbs.

So far as carburetor sizing is concerned, dual quads are not chosen in the same way as a single 4-barrel is. Ford installed a pair of 600ish cfm carbs on Shelby GT500s with a very mild hydraulic 428 engine and a warranty in 1968. There is certainly no reason to go any smaller today. A pair of 750 cfm carbs is perfectly acceptable on a more serious or larger-displacement build, even on the street.

Reproductions of the OEM dual quad carburetors are available from Holley and sold through Carl's Ford Parts. High-performance carburetors that are specifically built for FE dual quad installations are available from Quick Fuel Technology and sold through Survival Motorsports and a few other FE specialty distributors.

Either the original or available reproduction throttle linkage mounts to the rear carb and crosses in between the two carburetors to make the throttle connection.

The factory dual quad linkage is progressive: The front carburetor opens first; the secondary (rear) carb opens later but at a faster rate so that they reach wide open throttle at the same time. This system allows the car to run and drive as if it had a normal double pumper–style carburetor.

The aluminum oval air cleaner assembly atop the multiple carb system is an iconic FE visual item and is readily identified as "cool stuff" by any enthusiast.

The three 2-barrel setup is another vintage package, one that pre-dates the dual quad systems. Three 2-barrel combinations were offered on factory 390 and 406

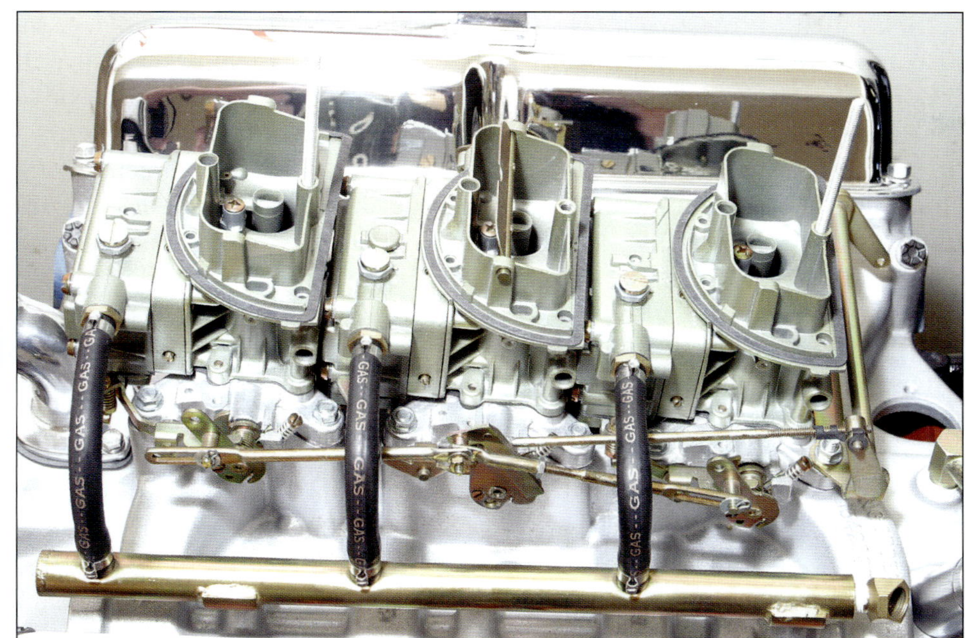

The three 2-barrel systems almost always use a factory Ford intake manifold. The Galaxie intake was angled, while the T-Bird one keeps the carbs flat/level to the engine. Like the dual quad, they use a progressive throttle linkage that starts the center carb first. The 3 x 2 package does not perform as well as the 2 x 4, or a good single 4-barrel. It is good-looking, though, and a viable option for those seeking that vintage cool factor in terms of cosmetics. (Photo courtesy Jim Smart)

The Carter M6905 mechanical fuel pump meets the needs of any street-oriented FE. It has the additional advantage of being cosmetically similar to the factory Cobra Jet fuel pump, and with its clockable fittings can be installed in virtually any FE-powered vehicle.

performance cars in the early 1960s. The three carb systems are visually appealing, although the dual 4-barrel setups will outpower them by a wide margin.

No reproductions of the original 3 x 2 manifolds were made, but performance aftermarket versions do exist. The manifolds for Thunderbird installations are flat front to back to accommodate the engine installation in those vehicles. Manifolds destined for full-sized cars, such as the Galaxie, have an angled carb mounting flange, higher in the rear, similar to most of the original 4-barrel and dual 4-barrel intakes.

The carbs for the 3 x 2 are unique in that the air cleaner flange is D-shaped, along with the expected calibration specialization. Some folks will adapt a common 2-barrel replacement Holley carb as a service replacement, but this involves significant modification.

Similar to the dual quads, the linkage for a triple carb combination is progressive, with the center carb starting first and the outer carbs being brought online further into the throttle action.

The mechanical fuel pump mounts to the timing cover on the driver's side of the engine. Use assembly lube on the lever arm and slide the arm below the pump eccentric inside the cover before pulling the pump up into position by hand to install the fasteners. Don't try to force the fasteners to pull the pump up; you'll strip the threads. Metal line from the pump to the carb with adequate support is mandatory to prevent problems down the road;

only a short connecting piece of rubber hose should ever be used for fuel under pressure.

While the stock replacement-style pumps are adequate for non-performance rebuilds, a high-performance fuel pump, such as the Carter M6905, should be considered for higher output street engines, along with 3/8-inch-minimum fuel lines. Ford used a very similar pump on the factory 428 Cobra Jet applications. A rear-mounted electric fuel pump is a good upgrade because the pressure throughout the fuel supply line will reduce the tendency for vapor lock and fuel boiling, a common occurrence with today's fuel chemistry.

A fuel filter on the pressure side of the system ahead of the carburetor is mandatory. The smallest amount of debris can cause the needles to stick and flooding to occur. Normal fuel pressure should never exceed 6.5 to 7.0 psi. Higher pressures will require improper float adjustments and can lead to tuning issues and flooding.

Carburetor Tuning

You often read about carburetor tuning issues in print and on the Internet. Many of the discussions are fraught with misinformation, overly expanded generalities, gross simplifications, and myth. While we could write a book on carb tuning (and several good ones do exist), I thought we should provide a little bit of guidance beyond the bolt-it-on basics.

First is a brief discussion regarding selection. For a long time, the enthusiast magazine community has emphasized using a small 4-barrel for the street. While this guidance is generally sound, keep in mind that Ford used +/- 600 cfm carbs on very mild stone-stock 390 engines back in the 1960s. There is no reason at all to consider anything smaller these days, and a larger vacuum secondary carb such as a 735, 750, or 780 will provide excellent street manners and more power in most performance applications. In a similar vein, dual quad packages are not sized in the same manner as single 4-barrels are. The factory dual quads on a 1967 Shelby GT500 KR were around 600 cfm; this on a hydraulic cam pump gas 428 sold with a warranty 50 years ago. Do not even consider using anything smaller on dual quads.

Vacuum secondary operation is perfectly fine, and even advantageous, on street applications. You utilize a variety of springs (or an adjustment screw on Quick Fuel carbs) to tailor the secondary opening rate to your needs. The double pumper with mechanical secondary operation certainly has its place on track vehicles and serious street/strip packages. The advantage of rapid mechanical secondary operation is not a very good match for heavy vehicles such as trucks or a big Galaxie, but it makes for a good package in a road race Cobra or stick-shifted Mustang.

Jetting is easy to change and equally easy to mess up, with potentially catastrophic results. The jet combination provided in a new carburetor is often on the rich (and safe) side. This is a good baseline to start from; do not immediately jump into changing things simply because you read about it somewhere. You need *data* before changing things, and data comes from a running engine using oxygen sensors, brake specific fuel curves, or simply reading plugs and evaluating driving characteristics.

The jet combination on a given carburetor is part of a package that includes the jets themselves, the power valve and power valve channel restrictions, the air bleeds, the booster and venturi design, the emulsion bleeds, and the engine demand. Nobody can tell you to "put 68 in the front and 72 in the back" without having a ton of additional information. A jet change that is more than four or five numbers away from the original calibration may well require additional changes to the bleeds to keep the fuel mixture reasonably linear at all RPM levels.

A stronger engine *does not* need bigger jets as a rule. The opposite is often true. An oversize carb on a mild engine *does not* require smaller jets; again the exact opposite may well be the case.

Power valves are another often misunderstood item. They are a vacuum operated valve that provides additional fuel under load as determined by manifold vacuum. The amount of fuel added to the overall mixture is determined by the diameter of two small passages in the metering block that are located behind the power valve with the descriptive name "power valve channel restrictions." These restrictions are additive to the fuel flowing through the jets and can be a significant percentage of overall fuel delivered to the engine.

The point at which the valve opens and additional fuel is provided is determined by the stamped number on the valve. A "6.5" power valve opens (or closes) when vacuum in the intake manifold reaches 6.5 inches as throttle and load increase. All the various Internet formulae for selecting power valves are simply to get you a safe starting point. A valve that opens too soon will cause a rich spot in part throttle, one that opens too late will cause a lean spot. Restrictions that are the wrong size will always be too rich or too lean, causing you to change jetting to compensate. The relationship between the jets and the power valve channel restrictions and the timing of the power valve opening point are all tuning tools to use if you are trying to optimize your system.

A power valve that has too large or too small a rating number will not cause idle issues at all, contrary to many old assumptions. The only time a properly functioning power valve might impact idle behavior is if it's right at idle vacuum on an engine where the vacuum value wobbles up and down, making the valve into a little pump as it opens and closes. An unlikely event. Fuel dripping from the boosters at idle is a sign of a different problem. The power valve is not causing it and changing the power valve will not fix it. Unless, of course, the valve has a damaged gasket or diaphragm; that's a whole 'nuther deal.

IGNITION AND DISTRIBUTOR

You have quite a few options when selecting your ignition system. You can run with a factory points distributor, you can convert the original distributor to electronic with a Pertronix kit (a factory electronic distributor is readily available), or you can go to an aftermarket distributor. Most original distributors, and several aftermarket ones, include a vacuum advance mechanism. Vacuum advance is highly desirable for part-throttle and steady-state cruise performance and gas mileage.

Any original distributor will need to be thoroughly disassembled, cleaned, and rebuilt. They are all 40 or 50 years old, with a lot of service life behind them. The advance mechanism has a pair of weights that work against springs to control the advance rate, and metal stops that limit how much mechanical advance you get. You can work with available kits to change the advance curve to suit your desires or you can send the distributor out to have it reworked by somebody with a distributor machine. Whatever you decide, resist the temptation to put the lightest and fastest possible curve in; its rarely the right answer for a heavy street car.

With the popular MSD distributors, the advance mechanism is easier to service since the springs and weights are located right beneath the rotor. The kit MSD supplies with every distributor includes an array of springs and bushings. The bushings control how much advance will be possible between base timing (usually assumed to be at idle) and maximum timing at a high RPM. The springs control how quickly the timing advances from the base to the maximum as RPM increases.

Original FE iron heads run best with total timing set somewhere between 36 and 40 degrees. Aftermarket aluminum heads are more efficient and perform best with somewhere between 34 and 38 degrees.

The factory Ford Duraspark distributor was used in 1975 and 1976 Ford pickups and is still readily available. It has a pickup coil and trigger inside, and works with an external ignition module. An excellent option if you wish to combine an original look with an aftermarket control such as an MSD 6AL, which can be hidden in the vehicle.

In the aftermarket, MSD distributors are very popular, although others are available and will certainly work well. MSD offers two versions for the FE. This is the basic PN 8594 billet distributor used on most race cars. It includes the pickup coil and is designed to trigger an external ignition box. No vacuum advance is included.

This MSD distributor is best suited to street use. The "Ready to Run" PN 8595 includes an electronic ignition module enclosed inside the distributor itself, along with a vacuum advance. Although this distributor costs more, it's actually a less expensive option once you factor in the cost of a separate ignition box.

The distributor gear must be compatible with your chosen camshaft. A flat tappet cam is made from cast-iron, and a cast-iron gear will work perfectly. A roller cam requires a gear that will work with its core material, often steel. While race cars will use bronze, that material will wear out quickly and is not desirable for long-term street use. A better choice is specially treated steel, such as this Crane item. These are available for the factory distributors, as well as for Mallory, PerTronix, and MSD products.

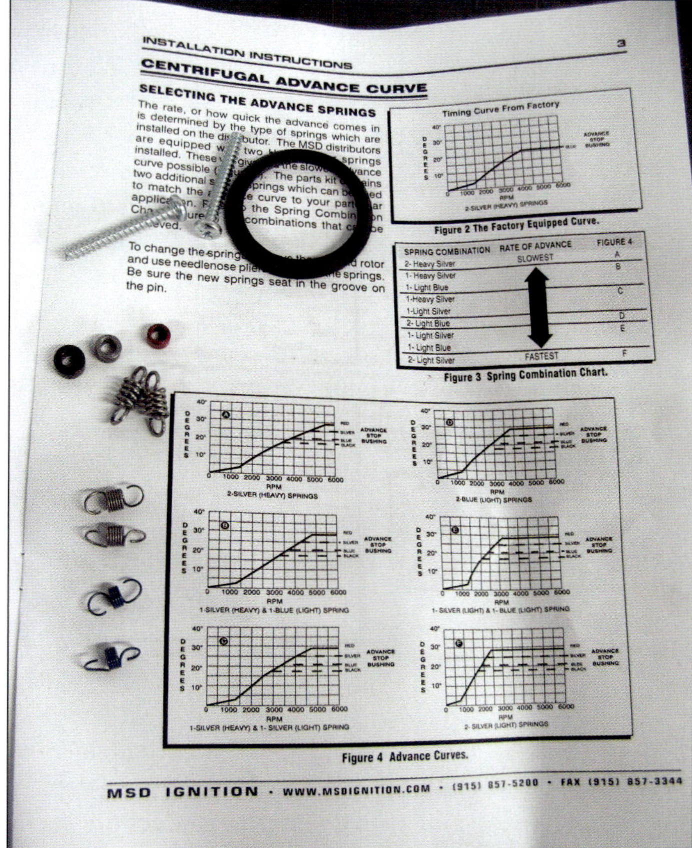

Figure 4 Advance Curves.

Both MSD distributors include a variety of springs and bushings to adjust the advance curve, which are accessible right below the rotor. Bushings determine the total advance between base timing and maximum timing. The springs determine how quickly that timing change occurs relative to engine RPM. As a rule, a larger car or one with a milder combination will want a slower curve than would a lightweight 4-speed car.

You also have the option of locking out the timing if you are building a dedicated race car or are using engine management software such as electronic fuel injection. Locking out timing requires removal of the distributor gear and collar. Remember to make this change, if desired, before installing any new gear or you will be doing it twice.

Some combinations will work best with less (or more), but an initial setting of 36 total will usually get you a safe place to start out. Initial timing can range anywhere from a factory single-digit value to somewhere around 16 or 18 degrees with a street/strip hot rod. Every combination is a bit different, and your cam and compression will be a major factor, as will your chosen fuel. Hot starts can become a problem when you are running a lot of initial timing. Locking out the timing can be viable for a race car but it's rarely the right answer on the street.

Perfect restoration folks might want to run the original points. The factory points system will work okay, but an electronic system of any sort is far superior. The Pertronix electronic conversion is inexpensive, bolts in place of the points in only a few minutes, and seems to work well on street vehicles. It allows you to retain the cosmetics of the original distributor while reaping the benefits of electronic ignition.

With any distributor you must be certain that the gear is compatible with your camshaft. Flat tappet cams work fine with the gear supplied on any distributor, but roller cams will require something different. We generally use a Crane steel gear on our street builds. The steel gear is compatible with all hydraulic roller cams. The only caveat is sizing; factory distributors have a .467-inch shaft, Mallory and Pertronix distributors have a .500-inch shaft, and MSD distributors have a .531-inch shaft. The .500-inch- and .531-inch-diameter gears are marked as fitting 351C and 460 engines, but they are perfectly interchangeable with the FE distributors. Distributor gears on Ford engines are press-on, and MSD

Visible from this angle is the proper position of the distributor relative to the intake manifold. The seal must rest below the manifold surface, the distributor housing slightly higher than the intake surface, and the clamp slightly angled upward and not bottomed out. If the distributor is much higher than this, you might not be fully seated or the oil pump shaft may not be properly engaged.

Spark plugs for the FE are tapered-seat 18-mm thread with a 1/2-inch reach when using factory iron heads. Aftermarket aluminum heads use a gasket seat 14 mm with a 3/4-inch reach. On street engines I normally start off with an Autolite 45 for iron heads or an Autolite 3924 on aluminum heads. I set gaps to .035 to start with and adjust heat ranges and gaps as I seek out the best performance over time. Plugs get a small amount of nickel anti-seize lubricant on the threads and get snugged up by hand.

provides a specific 3.045–3.050-inch target dimension from the bottom of the gear to the mounting index of the distributor body.

Installing the Distributor

To install the distributor, crank the engine around until you just feel compression at the spark plug hole on the passenger's side front number-1 cylinder. Keep rotating the engine until you reach +/- 20 degrees before top dead center on the damper. This will not be the perfect timing position for a running engine, but it will definitely be good for getting it to fire up quickly.

Note the location for number-1 spark plug wire on the distributor cap and transfer that location to the distributor body using a Sharpie. The tip of the rotor should point at that mark once the distributor is installed.

Lubricate the distributor gear, the seal, and the part of the housing that indexes into the block. Guide the distributor down into the hole in the manifold. You will want to lead the rotor a bit because it rotates a few degrees as it engages the gear on the camshaft. It is very likely that the oil pump drive will not slide right into the distributor shaft; you will need to rotate the engine slightly while you push or tap the distributor to get it to fully drop into place. When it's all the way down, the seal will be in

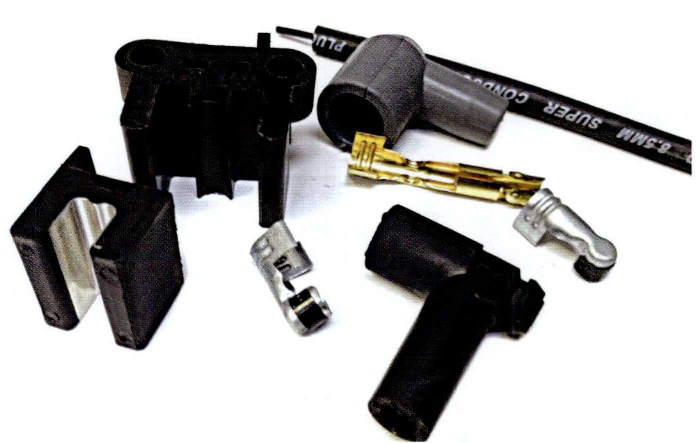

Plug wires should be a resistance-type wire and need to be compatible with the chosen distributor cap. My personal preference has been black MSD wire sets that we custom fit to each engine. Factory distributors and wire looms can accommodate only the smaller 7-mm wire size. Compatible with aftermarket distributors, the larger and arguably superior 8.5-mm wire size requires custom wire routing looms.

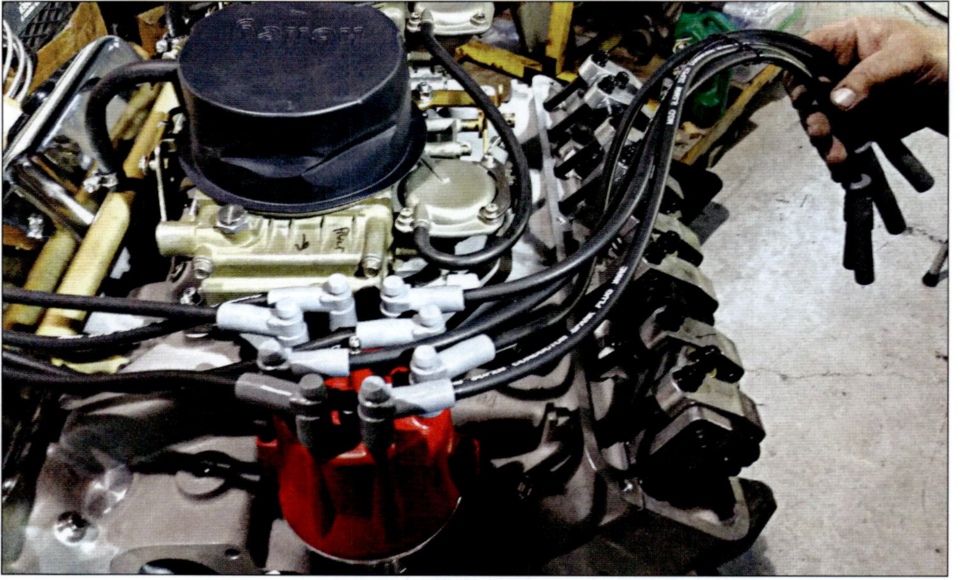

Both the cap and the plug end of each wire get a light coating of dielectric grease before installation. Starting with the number-1 plug wire, as previously marked on the cap, the wires are installed in the counterclockwise firing order of 1-5-4-2-6-3-7-8. Try to keep all wires away from linkages, and avoid coil and distributor electronic sensor wires. Also keep number-7 and -8 wires from running alongside each other in parallel; this can contribute to cross firing.

contact with the hole in the manifold, and the part of the distributor housing that the clamp contacts will be slightly above the intake manifold surface. Rotate the engine around a couple of turns to reset, verify the number of degrees BTDC timing location, and clamp it down.

Wiring is specific to the chosen distributor, but still merits a brief discussion. The factory points system will have a single wire going to the negative side of the coil. Easy.

A Pertronix points conversion will require an additional 12-volt key-on feed wire. It's often simply wired up to the positive-side coil wire, which works most of the time, but not always. The Pertronix system wants to see a true 12 volts but the points coil feed is through a resistor wire that effectively reduces the voltage by roughly a third. Sometimes this still works fine, other times it has been problematic. A secondary issue is that if you choose to bypass the resistor and go straight to the 12-volt side of the ignition switch, the optional factory tachometer will not like the additional voltage. If yours requires the straight 12 volts, you might consider using a relay for power and a resistor in the line heading back to the tach.

An MSD system with an external box (such as the popular MSD 6AL) also needs a key-on signal, but in experience seems somewhat indifferent to the actual voltage.

Pre-Oiling and Final Assembly

We got a little ahead of ourselves with the previous ignition chapter, as we should really pre-oil the engine before installing that distributor for keeps. This brief chapter covers the final assembly items and those lubrication processes needed before firing up our engine. Some of this might be redundant, but it bears repeating.

Oil Selection

An extensive range of ideas, data, opinions, and perhaps excess information is circulating about engine oils. I will reduce this to a few key concepts.

First is that, as good as synthetic oils may well be, it makes little sense to use a synthetic as an initial break-in lube for a stock or moderate street-performance engine build. Synthetics are sold on the premise that they last longer, give better protection at high temperatures, and pour and flow better at low temperatures. The first load of oil will be used only briefly and, since we will closely monitor the first minutes of operation, we will not see temperature extremes. Several years ago, I heard of ring sealing issues on break-in when running synthetics. While there is no way to truly prove, or disprove, those stories, not much of a case can be made in favor of running expensive synthetic oil for break-in.

Next is the use of a high-zinc oil. Over the past decade we have heard of a far greater number of flat tappet camshaft failures on initial break-in. The root cause of the reported increased failure rate has been blamed on everything from the

Most of the gaskets for the adapter mounts have a pair of simple holes for oil transit into and out of the filter. Take a moment to match the gasket's openings to those of the adapter with an X-Acto handicraft knife. I use a small amount of gasket sealer when attaching the mount to the block. This is a high–oil pressure connection and it will leak given any opportunity.

The oil filter adapter mount is an aluminum die casting mounted to the block at the driver's side front corner with four bolts. All common adapters have a pair of feed passages cast in place for oil going to and from the filter. Adapters made after 1968 had these passages enlarged and deepened for improved oil flow, and are preferred when the application allows. Aftermarket reproductions of these are available.

cam material to lifter manufacturing to cam design to inadequate break-in processes. One fact is that most motor oils have significantly reduced the amount of high-load additives, such as zinc and phosphorus, for emissions reasons. The market now carries a few specialty oils specifically formulated for break-in. From a risk versus reward perspective, it makes good sense to use one of these when firing up a flat tappet cam engine for the first time.

Last is that, for hydraulic roller engines in particular, the days of the really thick 20W50 oil seem to be gone. Lifter manufacturers prefer to see a 10W30 or 10W40 oil at the most. And plenty of engines, including almost every newer car engine for a decade, run perfectly well on a 5W-something oil. My preference as a break-in oil on a hydraulic roller engine is any brand-name oil in 10W30 that you can acquire at a good price. Remember, you will be recycling it after 30 or 60 minutes of run time.

Pre-Oiling the Engine

We are finally at the point where we can start getting ready to

I prefer Motorcraft or Wix oil filters. No real reason for that, just a lot of experience with no problems. Pre-fill the oil filter with your chosen oil, put a thin film on the gasket, and spin it into place.

While many causes have been listed for flat tappet cam journal failures during initial break-in over the years, a known fact is that everyday engine oils no longer have the zinc and phosphorus (additives that aided in break-in) that they used to. It's cheap insurance to use special break-in oil and a zinc additive to give your new cam and lifters the best chance for early survival.

To verify oil pressure, I use a common 0–100 psi pressure gauge available from any plumbing supply and thread it into the 1/4-inch NPT oil pressure fitting hole on top of the adapter. This is where the oil pressure sending unit will go on the completed engine.

Use a drill, a long extension, and a 1/4-inch-deep well socket to spin the oil pump. The drill will need to be able to run in reverse, or counterclockwise, for this step. Taping the socket to the extension is a good measure, ensuring you won't have to remove the pan again should the socket slip off the extension.

fire things up. The first item on the agenda is to check for oil flow and pressure. Pour the requisite amount of oil into the engine (I pour it down the distributor opening). Use a corded drill, a long extension, and a 1/4-inch-deep well socket as a priming tool. It is a really good idea to tape (or weld if you want a dedicated tool for future use) the socket to the extension so that it comes up when you remove the drill. I pack a bunch of paper towels alongside the valve spring side of the heads to catch the oil that is sure to run off.

Slip the socket and extension down onto the oil pump driveshaft and spin the drill counterclockwise. On most FE engines it will pick up oil pressure and really slow the drill after only a few moments. Rear sump pans with long pickup tubes will take a little longer. You should see between 60

This is a factory steel valve cover. Covers with this physical shape are often referred to as "pentroof" style. The Mercury logo shown is rather unusual, as most have no logo at all. Those on Ford performance engines were often chromed.

These finned aluminum Cobra 427 valve covers from Blue Thunder are a popular aftermarket accessory. Excellent casting quality and precise machine work are a Blue Thunder trademark.

The oil pump drive will be recessed into the hole in the bottom center of the distributor opening. Off to the side of that same opening you can just see the gallery plug location at the end of the driver-side lifter feed passage. If you inadvertently left that plug out, you would see a large stream of oil spray out from there, and you would remove parts to address the issue.

These are the factory aluminum valve covers used on many 428 Cobra Jet engines. Attractive die castings, they are also available as reproductions from Tony Branda if you cannot find originals.

Shown here is the iconic 427 "Bird" logo applied to a smooth chrome "baldy" valve cover as used in the 1963 and 1964 427 engines. These covers have no oil filler provisions and must therefore be used in conjunction with an intake manifold that incorporates an oil filler tube.

and 80 psi of oil pressure indicated on the gauge you have on the filter adapter. Once you have pressure you should start to see oil flow from each rocker arm after a minute or so of priming. Sometimes you might need to rotate the engine a couple times to get flow through all the lifters and rockers. In most cases you will have plenty of oil flowing within a couple minutes. If you do not have indications of oil flow at all the rockers, or if pressure is low, stop and figure out what is wrong.

Very low pressure can be caused by leaving a gallery plug out. While spinning the drill listen for a bubbling, splashing, or gurgling sound. The most common plugs to miss are the one behind the distributor and the one behind the upper timing sprocket. Either one will generate a big stream of unrestricted oil if left open, easily seen through the distributor hole or the fuel pump opening while priming.

No oil flow to rockers on one side will usually be the result of a tolerance stack-up that puts a rocker arm or cylinder head fastener right up against the oil feed passage, blocking the flow. Remove the fastener that borders the oil feed on the rockers and try priming again. If oil pours and gurgles to the top you have found the problem; just get that fastener's shank turned down on a lathe and continue on. Be sure

to remove any residual oil from the hole to prevent hydro-locking and subsequent cracks.

If you have no luck you'll need to remove the rocker assembly from the offending head and remove the nearby head bolt and again spin the drill. If it spits oil now, reduce that head bolt's shank diameter to the minor diameter of the threaded section. It won't hurt anything, and I have seen this happen a couple times over the years.

With oiling verified you can install that distributor as noted in the prior chapter.

Valve Covers

Installing the valve covers seems pretty simple, and it is. Even so, there are a few things to check. First is to make sure that they clear your chosen valvetrain. Virtually any cover will clear a stock rocker system. Aftermarket rocker systems, or the addition of end stand supports, can interfere with valve covers. The quickest way to spot check them is to "paint" the inside with red machinist's dye, set one on a head without a valve cover gasket, and wiggle it around a bit; you will immediately know if there is any contact, and the spot where it hits will show up. Sometimes a contact point will need just a couple hammer taps on a baffle. Other times it will be apparent that major mod-

ifications or alternative covers are needed. The thickness of the gasket will provide plenty of clearance once any interference has been addressed.

My preference is for cork gaskets. The rubber ones always seem to have a difficult time sealing the area between the intake manifold and the cylinder head, which is a unique FE issue. Put a bit of sealer at the junction of the intake manifold and the cylinder head. If the manifold gasket sticks up above the valve cover flange it's best to trim it level first. Many folks like to glue the gasket to the bottom of the valve cover. I generally leave them loose until the engine has been run and broken-in, since I often remove the covers for adjustment and inspection.

Valve cover fasteners should really have a lock washer or a serrated underhead. With a thick gasket under them, you do not want to tighten them up very hard. Just a hand snug is adequate, followed by another snug after the engine has run for a while.

If you have not already done so, you may now bolt the water pump into place. Some of the bolts go into the cooling passages, so a bit of Teflon paste is required. Remember to install the small bypass hose and clamps before putting the pump into position, as that hose is nearly impossible to replace afterward. Install the pulleys and verify belt alignment.

Finishing Up: Sensors, Thermostat, PCV, Pulleys, and Belts

The last parts include the sensors for oil pressure on top of the filter mounting adapter, and for water temperature in the threaded location behind the distributor on the intake.

We also have heater hose fittings (or plugs if not used) in the intake and water pump.

Install the thermostat in front of the intake manifold. The side of the thermostat with the spring and copper slug goes in toward the intake. If you are running the engine on a dyno for break-in, you will usually install the thermostat afterward. Engines from 1965 and earlier have a comparatively large-diameter thermostat and matching housing. Later models have a more readily available small-diameter thermostat and housing. Original thermostat housings were aluminum and were prone to cracking from misalignment in service. Many had additional tapped openings for emission or temperature/vacuum switches and senders. Aftermarket housings are generally more durable, albeit less attractive, cast iron. The fasteners for the thermostat sometimes go into water, so a dab of Teflon sealant is a good idea.

A positive crankcase ventilation system is highly recommended for any street engine. These use a valve connected to an intake manifold vacuum in one valve cover, and a breather in the opposite valve cover. You need both for the system to work: Vacuum at idle and cruise will pull crankcase vapors and moisture out from one side through the PCV valve, and fresh "make-up" air is introduced through the opposite cover's breather. The PCV helps reduce moisture damage from inside the engine and reduces the potential for leaks and oil mist in the underhood compartment. If you are running the 1964 or earlier valve covers with no openings, you can still make a functional, although less effective, PCV system using the rear intake breather opening originally intended for a road draft tube.

The front-end dress on the engine is often left off until after installation into the car, but it is still well worth test fitting everything. Pulleys on the front of the engine for water pump, alternator, and crankshaft should be checked for alignment and straightness. They sometimes get bent during removal, and if alignment is off, the engine will throw belts at higher RPM. Some replacement water pumps do not have the threaded holes located correctly for the various mounting brackets. All of these things are easy to fix on the engine stand and difficult to address or even find once installed into the vehicle. If dyno testing, we install a short belt to spin just the water pump.

The dipstick tube is usually installed once the engine is in the car. It taps/presses into the hole in the block, and the upper tab gets bolted to either the front of the cylinder head or one of the exhaust manifold bolts. These tend to get loose and leak; a bit of TA-31 silicone on the outside of the tube helps. If you choose to use an oil pan–mounted dipstick on a truck or a Cobra, you will need to close off the one in the block. A small piece of aluminum or an appropriate size oil gallery plug does the trick.

This is a large-diameter thermostat housing. They were common up through the middle 1960s and use the same thermostat found in numerous Chrysler applications. The water reservoir tanks used in Galaxies have the same size thermostat opening.

This is the much more common smaller-diameter thermostat housing. Aftermarket replacements are readily available both as generic iron parts and as aluminum reproductions for restoration efforts. It uses the same thermostat as virtually every other Ford V-8 from the 1970s on. Be careful when installing these: It's easy to get the thermostat out of position and crack the housing during fastener tightening.

EXHAUST

The engines installed in the majority of full-size passenger cars and trucks used a common 8-bolt exhaust mounting pattern. Even though they will physically bolt right on, not all manifolds or headers are compatible with all the cylinder heads due to port exit variances. Much better to check and address that now rather than spend lots of time figuring out persistent exhaust leaks later on.

The GT390 engine used a unique 14-bolt exhaust pattern that is not common to any other FE application. Headers for that pattern are available, but they are not interchangeable with other heads and will not work on Edelbrock heads if changes are planned in the future.

The 428 Cobra Jet heads use a 16-bolt pattern that includes the eight up-and-down locations common to traditional FE heads, allowing a degree of interchangeability with headers. The aforementioned cautions regarding port location apply. The Edelbrock heads and the 428 CJ heads share a common bolt pattern and port location.

On some vehicles you can install the exhaust manifolds or headers before putting the engine in; on others you mount them up only after setting the engine into the bay. Experience and test fitting is the only way to know.

Exhaust manifolds were usually installed with no gaskets from Ford. For that to work you need to have perfectly clean, flat, and smooth surfaces on both the head and the manifold. Any warp, debris, or damage to the surfaces is a guaranteed leak.

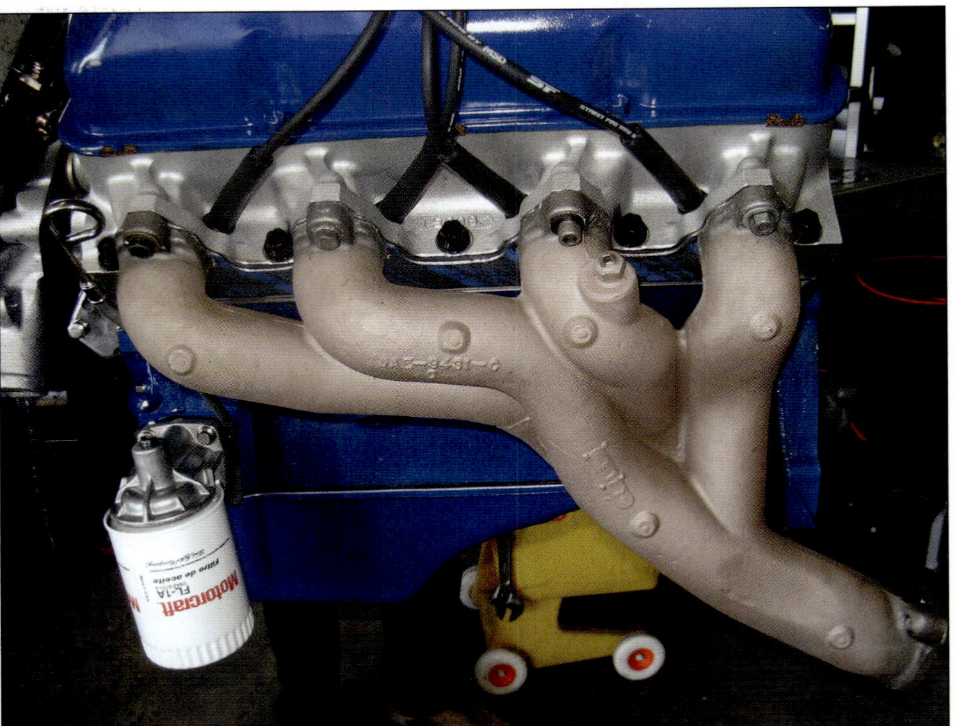

This illustrates a very rare set of 427 long tube manifolds, as well as the most common 8-bolt up-and-down exhaust fastener configuration for FE exhaust manifolds. Variations to accommodate FE engines in intermediate sized chassis with shock towers exist as well. Always make sure to check port and fastener configuration before installing the assembled engine in your car. Exhaust manifold removal from an FE engine in a shock tower car is not for the faint of heart.

These are the original 428 Cobra Jet exhaust manifolds. Of note here are the unique side-to-side fastener locations. The 428 CJ heads, along with most popular aftermarket aluminum heads, will have 16 threaded holes to accommodate both the up-and-down as well as the side-to-side fastener locations. Most headers will include both sets of bolt holes.

If you are using headers, or are not confident in the manifold's surface, you will need a header gasket. I have good luck with Fel-Pro or similar header gaskets that have a perforated metal inner layer. The plain paper gaskets have always failed on me at some point, even if they were initially sealed up. I always use a very thin layer of high-temperature copper silicone on both sides of the gasket.

Any manifold or header should have the bolts installed with a bit of anti-seize compound. If you are using stainless-steel fasteners into aluminum heads this becomes mandatory; otherwise, they will gall up and seize very quickly. Small hex header bolts make the task easier, but putting headers on an FE-powered Mustang or Fairlane is always going to be a chore, with limited room and no visibility.

Stand back and admire your completed engine!

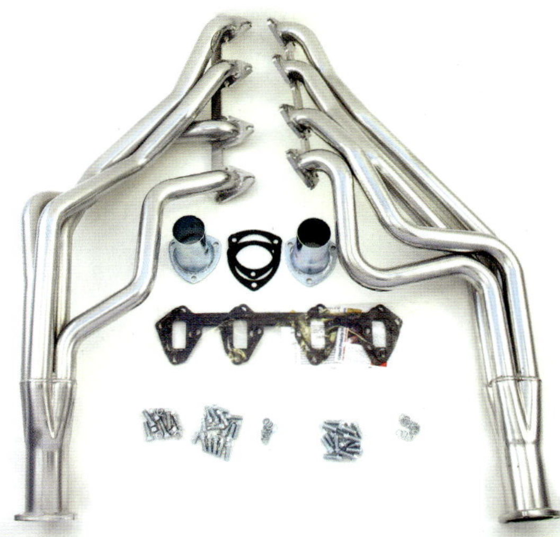

With the notable exception of the rare 427 long tube parts, the cast-iron manifolds used on most FE applications range from marginal to horrible in terms of flow. A set of headers will yield major improvements, especially on vehicles originally equipped with the log manifolds that are very tight to the engine.

Shown here is the completed 428 Cobra Jet that was the primary focus of this book, all finished up and ready to be run on the dyno. Make note of the special dyno headers, which would not fit in a car but are needed to fit the test machine. The lines running into each primary header tube are thermocouples to measure exhaust gas temperature for each cylinder. Farther back are oxygen sensors in the collector, which are used to monitor air/fuel ratios.

And here is the completed 428 Cobra Jet engine installed in the 1969 Shelby GT500 that it originally came from. A thing of beauty for any FE fan!

INITIAL START UP AND BREAK-IN

We normally fire up and test our engines on a dyno, but the same general processes are followed if you are doing yours in a car or truck. As you would expect, the key piece of advice is to proceed slowly and carefully, and don't get overly excited or rushed for time. If you need a few more minutes, or a couple more parts, wait until you can get things done right and have the time to go about the process at a comfortable pace. I have seen, and caused, a fair amount of trouble by getting ahead of myself at the very end of a project. YouTube is full of great videos of folks who destroyed their projects by getting into a rushed panic at the end.

Hook up all the required coolant hoses, and if on a dyno, block off the ones you will not need, such as heater connections. Check to be certain that the block drains are both installed. Fill the engine and radiator with water and look for leaks. Tighten clamps and seal plugs and bolts as needed before moving forward.

Once you are confident that the engine is going to be adequately cooled, we will move on to the wiring. All the ignition wiring needs to be tight, professionally terminated, and routed away from exhaust and fuel whether in car or on dyno. Even temporary wiring can cause lots of trouble if it gets burned through and shorts out. This is not the time to have stuff twisted together and held with tape, vise grips, and plans to tag the battery post for cranking and pull a wire to turn it off.

Check the carburetor and linkage for wide open throttle and throttle return to idle before adding fuel. It must be smooth operation with no tightness or sticky spots. Any interference or problems must be addressed now. Having your new engine stuck at wide-open throttle is dangerous and destructive to your pride and your checkbook.

Hook up the fuel lines as needed. With no power to the ignition, check for fuel leaks. On an electric fuel pump or on the dyno this is easy: Momentarily hit the pump switch and watch. If you see fuel leaking

Whether you are doing your initial start-up and break-in on the dyno or in your engine bay, a number of fundamentals must be checked before getting started. Make sure everything that can leak is tight and secure. Double-check and make sure throttle linkage is operating smoothly and not binding. If you have done restoration work other than the engine, prime the system and make sure fuel is getting to the carb, either via a mechanical or electric fuel pump. After checking everything once, go through the list again. Also, have a friend and a fire extinguisher nearby, and make sure you have the means to keep the engine cool for approximately a 20-minute break-in period.

at a fitting or see it running into the engine through the carb's boosters, stop and fix the problem now. Fuel pressure should be under 7 psi, and a leaking needle or stuck float can wreak havoc on a new engine fire up.

You probably need to wait until the engine is running to check the mechanical pump. Once you are as confident as you can be that everything is tight and leak free, you can add some fuel to the carb using a small funnel through the bowl vents (an electric pump will have already done this). Push the accelerator pump a couple times to verify that fuel is getting into the engine.

An assistant is really useful at this point. He can take the pictures and videos while you keep your concentration on the engine and start-up process. He can also help watch for issues and grab the things you need as the engine gets running. Have a fire extinguisher nearby just in case. If the engine is installed in the car, a garden hose can be handy to spray the radiator and keep it cool during break-in. I leave a timing light hooked up and ready to use.

Crank it over a few times with the ignition off to see that oil pressure comes up. Give it a couple shots of fuel by opening and closing the throttle. Turn the ignition on and fire it up. If you get a backfire through the carburetor, the odds are that the ignition timing is a bit retarded; advance it and try again. If it cranks over several times with no indication of starting, you should first check for spark. The easy way is to clamp a timing light to the coil wire and crank the engine; it should flash frequently. If you have spark, and you have fuel, and the timing is anywhere close to right, it's going to run. New engines are scary and fun. They smoke as the assembly lube and handprints burn off; they smell like oil and hot paint; they tend to burp and fart as they light off for the first time.

A lot has to be checked and verified in the first few moments of a new engine's life. Keep calm and work your way through the list. Particularly on flat tappet cams you should try to get the engine up to roughly 2,000 rpm as soon as possible and keep it there for about 20 minutes or so. The oil splash aids in the cam break-in process. Once running you will also want to check for leaks and listen for noises. It is perfectly okay to shut the engine off to fix things, and much safer than trying to work through any issues that arise while it is running. Check the ignition timing to be certain it is at a reasonable level for running. Thirty degrees of timing at 2,000 rpm is perfectly fine; you can get it to the best amount later on after things are stabilized and you are entering the tuning phase. You can move the throttle around a bit and vary the RPM but don't let it idle if doing a flat tappet cam break-in. If you removed the inner valve springs for flat tappet break-in do not rev it up very high either. Keep an eye on coolant temperatures, especially if doing this in the car. Running for a long time without adequate airflow can get things pretty warm, so an occasional light spray across the radiator from the garden hose can help.

After the 20 minutes of running time you can turn it off, grab your favorite cold beverage, and relax for a moment. You will need to install the inner valve springs if they were left out for break-in purposes. In any case, remove the valve covers to verify lash or preload, and check for any issues. This is also a good time to inspect spark plugs and go over all the wiring and hose connections to make certain that it's all clean, tight, and leak-free. Put a wrench on all the external fasteners for timing cover, oil pan, and such, as they tend to loosen a bit after heat and vibration from the initial running. Remove the oil filter, drain it, and cut it open, spreading the paper element out and looking for debris. A few tiny flakes and very small bits are normal in a fresh build, but any large amount of trash or metal in the filter is cause for alarm and investigation before things get critical.

Install your inner valve springs, a new filter, and a quart of oil and you are ready to start making power if on the dyno, or you are ready for some testing and tuning in the car. After restarting you will need to set ignition

Critical Inspection

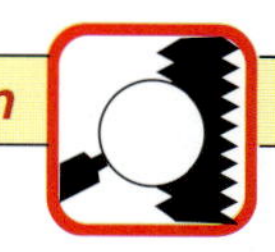 *After the initial break-in run, let the engine cool. Remove the oil filter, cut it open, and inspect it to see what is in the filter element. Small particulate matter is normal, but large amounts of trash or metal merit further investigation before doing any more harm.*

If you are running dual valve springs, and were only running the outer springs for cam break-in, now is the time to install the inner springs. Once installed, you can install a new filter and fresh oil, and get ready to start it back up for some fine-tuning. This is the stage where you will be setting the timing and fine-tuning the idle.

timing to a safe running and driving setting. My experience is that 36–38 degrees total timing (with vacuum advance disconnected) is a solid baseline to work from. Get the carb set to a safe and consistent idle with mixture screws somewhere between 3/4 and 1½ turns out from closed.

Put some moderate load on the engine. The piston rings will require cylinder pressure to seat, so it's best to get the engine "working" fairly soon. On the dyno we would be making pulls from 3,000 to 5,000 rpm; in the car you want to make a series of short, moderate accelerations in second or third gear. As you accumulate time and confidence in the build, you can tune the fuel mixture and timing to get the best performance by looking at the spark plugs, looking at data, or running down the track.

Once it has been broken in and everything is verified as healthy, there is no need to baby the engine for any period of time. I put a fresh engine under load and have it working as soon as I can on the dyno. Well-machined and well-assembled engines will be completely settled in and ready to go very quickly. I see them starting to repeat power numbers after four or five pulls on the dyno.

The first oil change should come pretty quickly, after no more than a few hours of run time. The initial oil you drain out will probably be dark and have a shimmering appearance. This is normal and should not cause alarm. More important is to open up the oil filter and inspect the media again to check for any debris. The fresh oil should be appropriate for the chosen cam design. By now things should be stabilized as far as valve adjustments, timing, and mixture, and you should be able to enjoy your engine for a long, long time.

```
                      Survival Motorsports
                     4202 Pioneer Drive, ste e
                      Commerce Twp,. MI 48390

Date : 10-31-2016       Time : 13:36:54
INF=1.28       Mode : A-Sweep

T14:15 D06-24                          jets p 63 (-3) s 75 (-4)


FileName : PANKOW69.D07
```

Speed RPM	CPower C_HP	C_TQ lbft	OIL PSI	A/F R right	A/F L left	Water Deg F	Fuel PSI	CIA F°
*3000	257.9	451.5	66.8	12.8	12.77	136	7.5	77
*3100	269.2	456.1	67.4	12.81	12.79	136	7.5	77
*3200	279.4	458.6	67.8	12.69	12.81	136	7.5	77
*3300	287.5	457.5	68.2	12.59	12.73	136	7.5	77
*3400	296.7	458.3	68.6	12.58	12.81	136	7.4	77
*3500	306.7	460.3	69	12.61	12.85	136	7.4	77
*3600	316.6	461.9	68.9	12.71	12.93	136	7.4	77
*3700	326	462.7	68.9	12.84	12.99	136	7.3	77
*3800	334	461.6	69	12.96	12.93	136	7.2	77
*3900	340	457.9	69.4	13.06	12.89	136	7.2	77
*4000	346.6	455	69.7	13.08	12.89	136	7.1	77
*4100	352.3	451.3	70.1	13.09	12.94	136	7.2	77
*4200	357.8	447.4	70.4	13.14	13.03	136	7.2	77
*4300	363.1	443.5	70.7	13.23	13.08	136	7.1	77
*4400	368.1	439.4	70.9	13.26	13.09	136	7.1	77
*4500	373.4	435.9	71.4	13.18	13.11	136	7	77
*4600	376.9	430.4	71.7	13.17	13.05	136	7	77
*4700	379.7	424.3	72.3	13.2	13.01	136	6.9	77
*4800	383.1	419.2	72.5	13.22	13	136	6.9	77
*4900	386	413.7	72.8	13.28	12.94	136	6.9	77
*5000	388.5	408.1	73.3	13.39	12.98	136	6.9	77
*5100	388.7	400.3	73.7	13.36	13.1	136	6.9	77
*5200	389.4	393.3	74.3	13.38	13.14	136	6.9	77
*5300	389.2	385.7	74.6	13.37	13.16	136	6.8	77
*5400	387.9	377.3	74.8	13.34	13.28	136	6.8	77
*5500	387.3	369.8	75.5	13.3	13.28	136	6.8	77
Average data in * band								
4250	347.43	433.93	70.87	13.06	12.98	136	7.13	77

This is a dyno sheet of the output from a test session. The dynamometer is a great device for both break-in purposes and validation of a new build's integrity. It lets you determine the engine's best performance based on timing and carburetor jetting changes. Most folks focus on just the peak horsepower and torque numbers, but a lot more great information is generated as well. On this printout we can see oil pressure and air-fuel ratio per data from oxygen sensors mounted in the header collectors. The commonly accepted best range for air-to-fuel ratios on pump gas engines is between 12.6 and 13.2:1 at wide-open throttle.

There is nothing like the look of a freshly rebuilt FE engine sitting in the engine bay, ready to provide miles and miles of exhilarating and reliable horsepower. This is an image of a job well done.

ARP
1863 Eastman Avenue
Ventura, CA 93003
800-826-3045
arpfasteners.com

ATI Performance Products
6747 Whitestone Road
Baltimore, MD 21207
877-298-5039
atiracing.com

Billet Specialties
500 Shawmut Ave.
La Grange, IL 60526
800-245-5382
billetspecialties

Blue Thunder
255 North El Cielo, Suite 140-499
Palm Springs, CA 92262
760-328-9259
bluethunderauto.com

Canton Products
232 Branford Rd
Branford, CT 06471
203-481-9460
cantonracingproducts.com

Clevite Engine Parts
1350 Eisenhower Place
Ann Arbor, MI 48108
800-338-8786
clevite.com

Cloyes Gear & Products
7800 Ball Road
Fort Smith, AR 72908
479-646-4662
cloyes.com

Competition Cams
3406 Democrat Rd
Memphis, TN 38118
901-795-2400
compcams.com

Danny Bee Racing
30752 Imperial Street
Shafter, CA 93263
661-746-0517
tcwperformance.com/

Diamond Pistons
23003 Diamond Drive
Clinton Twp, MI 48305
877-552-2112
diamondracing.net

Doug Thorley
1180 Railroad Street
Corona, CA 92882
800-347-8664
Dougthorleyheaders.com

Dove Manufacturing
27100 Royalton Road
Columbia Station, OH 44028
440-236-5139
doveengineparts.com

DSC Motorsports
59748 Reynolds Way
Anza, CA 92539
951-763-9765
dscmotorsport.com

Edelbrock
2700 California Street
Torrence, CA 90503
310-781-2222
edelbrock.com

Erson
16A Kit Kat Drive
Carson City, NV 89706
800-641-7920
pbmperformance.com

Federal-Mogul (Sealed Power,
Fel-Pro)
2555 Northwestern Highway
Southfield, MI 48075
248-354-7700
federal-mogul.com

FPA Headers
2526 23rd Avenue
Puyallup, WA 98371
253-848-9503
fordpowertrain.com

Holley Performance
1801 Russellville Road
Bowling Green, KY 42101
270-781-9741
holley.com

Hooker
704 Highway 25 South
Aberdeen, MS 39730
270-781-9741
holley.com

Manley Performance Products
1960 Swarthmore Avenue
Lakewood, NJ 08701
732-905-3010
manleyperformance.com

Melling Oil Pumps
P.O. Box 1188
Jackson, MI 49204
517-787-8172
melling.com

Milodon
2250 Agate Court
Simi Valley, CA 93065
805-577-5950
milodon.com

Moroso
80 Carter Drive
Guilford, CT 06437
203-458-0542
moroso.com

MSD Ignition
1350 Pullman Drive, Dock #14
El Paso, TX 79936
915-855-7123
msdignition.com

Precison Oil Pumps
2324 Decatur Avenue
Clovis, CA 93611
559-325-3553
precisionoilpumps.com

Quick Fuel Technoogy
129 Dishman Lane
Bowling Green, KY 42101
270-793-0900
quickfueltechnology.com

Robert Pond Motorsports
1651 Coteau Drive
Riverside, CA 92504
909-376-2530
robertpondmotorsports.com

Scat Enterprises
1400 Kingsdale Avenue
Redondo Beach, CA 90278
310-370-5501
scatenterprises.com

Shelby Enterprises
19021 S. Figueroa Street
Gardena, CA 90248
310-538-2914
carrollshelbyenginecompany.com

Survival Motorsports
4202 Pioneer Drive, Suite E
Commerce, MI 48390
248-366-3309
survivalmotorsports.com

T&D Machine
4859 Convair Drive
Carson City, NV 89706
775-884-2292
tdmach.com

Total Seal
22642 North 15th Avenue
Phoenix, AZ 85027
623-587-7400
totalseal.com

Trend Performance
23444 Schoeherr
Warren, MI 48089
800-326-8368
trendperform.com

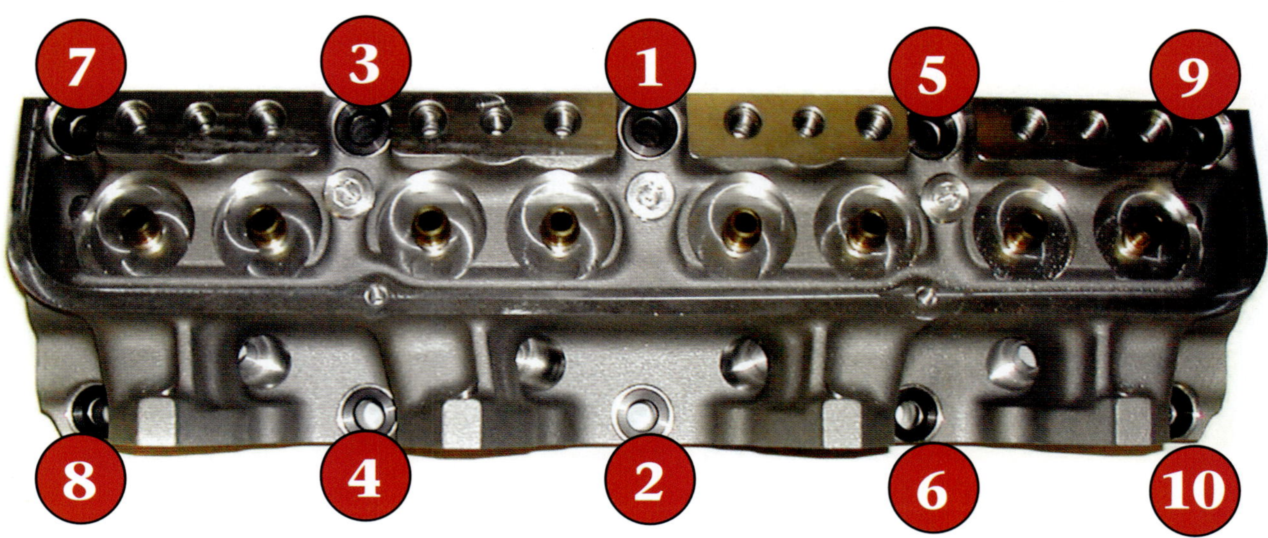

Cylinder Head Torque Sequence

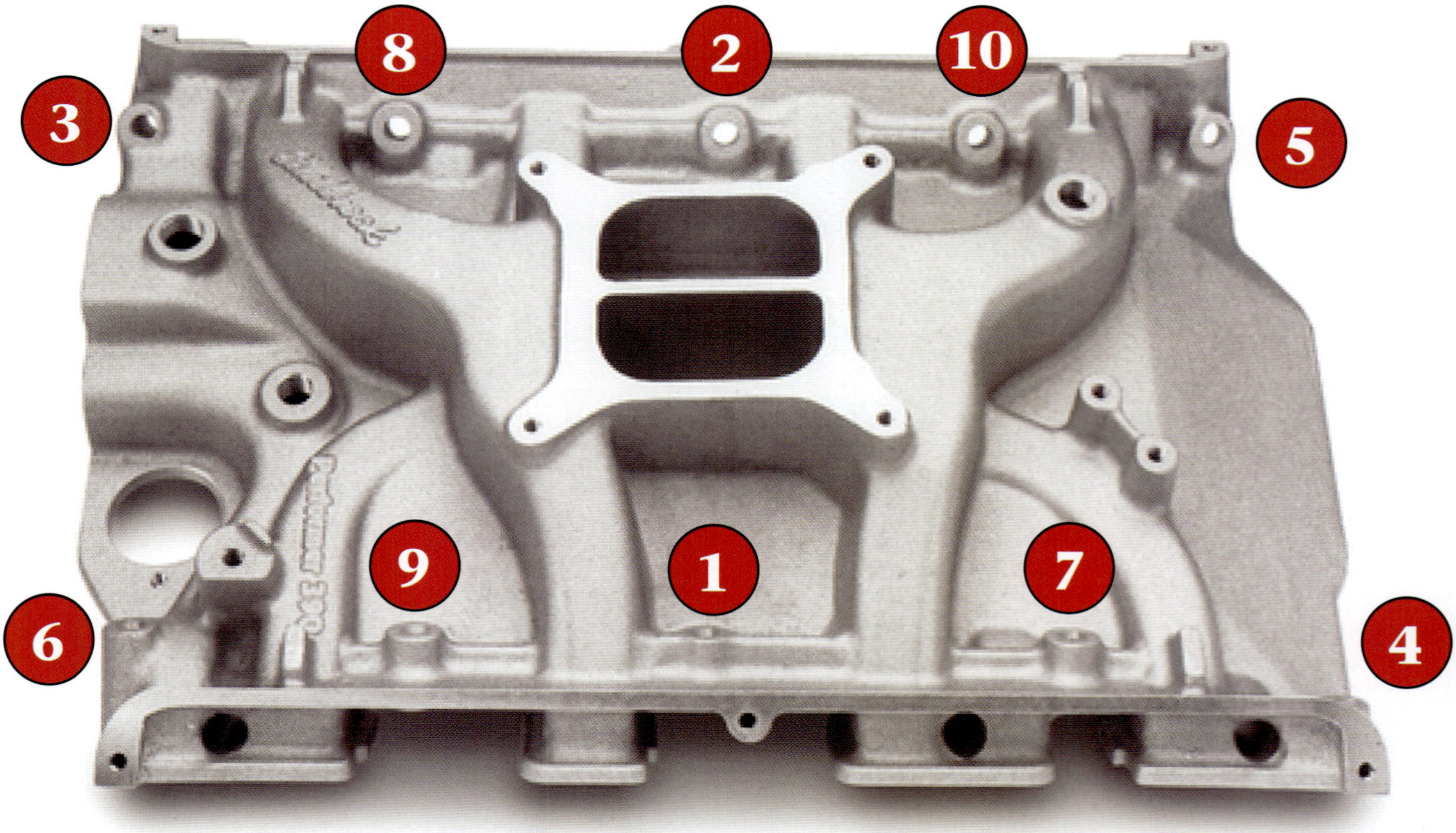

**Intake Manifold Torque Sequence
(Start all bolts before tightening!)**

Main Bearing Cap Torque Sequence

Main Bearing Cap Torque Procedure

1. Hand thread all fasteners.
2. Torque main bearing caps starting with number 2, then number 4, then number 1 in two stages.
3. Set the thrust bearing by knocking the crank toward the back, then toward the front with a soft mallet.
4. Torque main number 3 in two stages.
5. Torque main number 5 in two stages, following seal installation instructions in chapter 3.
6. If you are torqueing cross bolt mains for 406/427 blocks, torque all the mains first, following the procedure above, then torque the cross bolts starting at number 3, followed by number 2 and finally number 4.

Factory Torque Specs

Bolts or Parts	Lube or Sealer	Torque to:
Main Caps	Engine Oil	105 ft-lbs.
Main Cap Crossbolts (406 and 427)	Engine Oil	40 ft-lbs.
Warning!!! If you are using ARP bolts, you MUST use their specs.		
Connecting Rod Bolts 427 and 428 SCJ LeMans	Engine Oil	45 ft-lbs. *55 ft-lbs.
Cylinder Heads *1963-1967 427	Engine oil (blind hole) Sealer (water jacket)	90 ft-lbs. *110 ft-lbs.
Rocker Arms	Engine Oil	45 ft-lbs.
Oil Pump	Engine Oil	35 ft-lbs.
Oil Pickup with Threader	Engine Oil	15 ft-lbs.
Oil Pan	Engine Oil	12 ft-lbs.
Cam Bolt (upper gear)	Thread Locker	50 ft-lbs.
Front Cover	Engine Oil	20 ft-lbs.
Intake Manifold	Non- Hardening Sealer	35 ft-lbs.
Valve Cover	Engine Oil	10 ft-lbs.
Flexplate (Automatic) Flywheel (Clutch)	Thread Locker	85 ft-lbs.
Clutch Pressure Plate	Thread Locker	35 ft-lbs.
Centerbolt (Harmonic Damper)	Thread Locker	90 ft-lbs.
Bellhousing (Transmission to Block)	Engine Oil	50 ft-lbs.

Firing Order 1-5-4-2-6-3-7-8

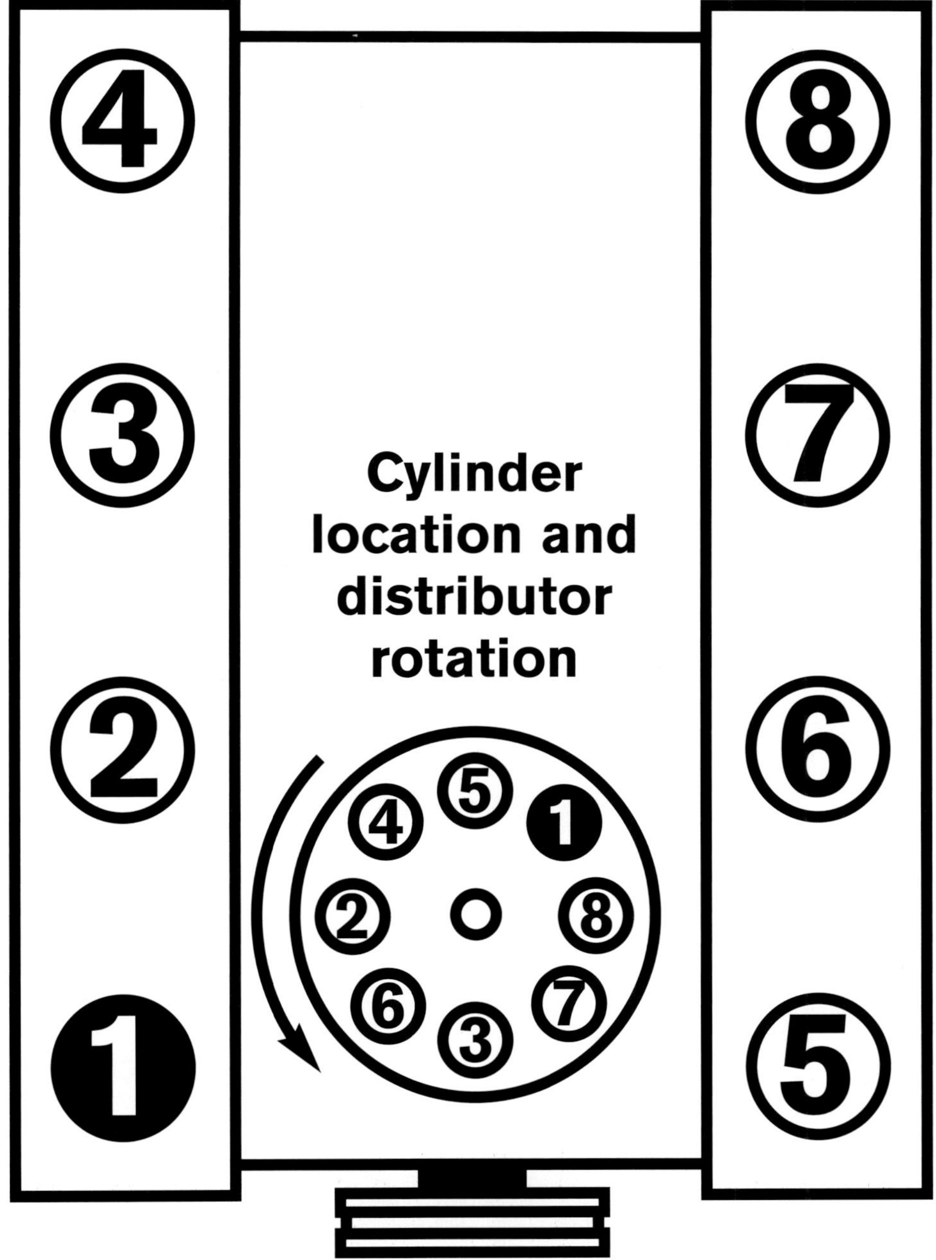

S-A Design *Work-A-Long Sheet*®

DISASSEMBLY

Project Statistics

Your Name ___________________________

Today's Date ______________ Vehicle Engine Removed From ______________

Engine Year ______________ CI ________ Block Casting ____________ ☐ 2 barrel ☐ 4 barrel ☐ Fuel Injection

Accessories Attached to Used Engine

☐ A/C Pump   ☐ AIR Pump   ☐ AIR Distributor Lines and Hoses   ☐ Water Pump
☐ Flywheel   ☐ Clutch   ☐ Flexplate   ☐ Transmission
☐ Starter   ☐ Fuel Pump   ☐ Exhaust Manifolds ☐ All Pulleys; Except ______________
☐ Alternator   ☐ Distributor   ☐ Coil   ☐ Carburetor
☐ Motor Mounts   ☐ Motor Mount Attaching Brackets   ☐ Spark Plug Heat Shields
☐ EGR Valve   ☐ Dipstick Tube   ☐ All Bolts; except ______________
☐ ______________   ☐ ______________   ☐ ______________   ☐ ______________   ☐ ______________

Operational Notes

Oil consumption ______________   Compression check pressure variation ______________ psi
Leak-down percent ______________   Other observations ______________

Disassembly Notations

Crank uses centerbolt   ☐ Yes   ☐ No
Heat riser restricted on ☐ Left   ☐ Right   ☐ Both
Head gaskets   ☐ Steel shim ☐ Composition
Worn/damaged lifters   ☐ No   ☐ Yes;   where ______________

Vibration damper pulley screws   ☐ 3/8-NC   ☐ 3/8-NF
Location of timing-pointer attaching points:
Oil filter adapter type:
☐ Spin-on
☐ Long cartridge (late)
☐ Short cartridge (early)
Type of rear main seal:
☐ Rubber—two piece
☐ Rubber—one piece (late)
☐ Rope (early)

INSPECTION

Initial Parts Inspection Observations

Block OK   ☐ Yes   ☐ No; describe problem ______________
Heads OK   ☐ Yes   ☐ No; describe problem ______________
Crank Ok   ☐ Yes   ☐ No; describe problem ______________
Bearings OK   ☐ Yes   ☐ No; describe problem ______________
Pistons OK   ☐ Yes   ☐ No; describe problem ______________
Cam/lifters OK   ☐ Yes   ☐ No; describe problem ______________
Damper OK   ☐ Yes   ☐ No; describe problem ______________
Intake manifold OK   ☐ Yes   ☐ No; describe problem ______________
Exhaust manifold OK ☐ Yes   ☐ No; describe problem ______________
Oil pump OK   ☐ Yes   ☐ No; describe problem ______________

Oil pump/rear main cap mating surfaces damage/abnormalities   ☐ No   ☐ Yes

Identifying mark you placed on all parts:

____

AT THE MACHINE SHOP

Parts Delivered to the Machine Shop

☐ Block	☐ Main Caps	☐ Crankshaft	☐ Oil Pump	☐ Oil Pump Pickup
☐ Connecting Rods	☐ Pistons	☐ Piston Rings	☐ Camshaft	☐ Lifters
☐ Vibration Damper	☐ Main Bearings	☐ Rod Bearings	☐ Cam Bearings	☐ Rod Bolts
☐ Gasket Set	☐ Push Rods	☐ Rockerarms	☐ Head Bolts	☐ Main Bolts/Studs
☐ Miscellaneous Nuts/Bolts/Brackets for Cleaning			☐ _______	
☐ Water Pump	☐ Timing Cover	☐ Oil Pan	☐ Flywheel/Flexplate	
☐ Clutch	☐ Exhaust Manifolds	☐ Motor Mounts	☐ Motor Mount Attaching Brackets	

☐ Assembled Heads   ☐ Disassembled Heads with: ☐ Valves   ☐ Springs   ☐ Retainers   ☐ Keepers

☐ _______   ☐ _______   ☐ Rocker Balls and Nuts   ☐

☐ Intake Manifold   ☐ With Heat Riser Shield   ☐ Installed   ☐ Not Installed

☐ _______	☐ _______	☐ _______	☐ _______	☐ _______
☐ _______	☐ _______	☐ _______	☐ _______	☐ _______
☐ _______	☐ _______	☐ _______	☐ _______	☐ _______

☐ Other Accessories ____

Special Instructions for Machine Shop

☐ Bore block   ☐ Use torque plates   ☐ Desired piston-to-wall clearance: 0. _______ -inch

☐ Grind crank   ☐ Rod bearing clearance: 0. _______ -inch   ☐ Main bearing clearance: 0. _______ -inch

☐ Deck to clean   ☐ Surface heads   ☐ Install cam bearings   ☐ _______

☐ _______________________   ☐ _______________________

☐ _______________________   ☐ _______________________

Is pilot bushing to be installed in crankshaft (required for manual transmission)?   ☐ Yes   ☐ No

Are intake manifold heat shield holes to be tapped for 8-32 screws?   ☐ Yes   ☐ No

After You Pick Up Your Parts

☐ Yes ☐ No	Threaded holes reconditioned/chased	☐ Yes ☐ No	Drilled holes and edges chamfered
☐ Yes ☐ No	head/block dowels properly installed	☐ Yes ☐ No	Are cam bearings properly installed
☐ Yes ☐ No	Galleries tapped for screw-in plugs	☐ Yes ☐ No	Add 0.030-inch hole in gallery plug
☐ Yes ☐ No	Add 0.030-inch hole in thrust face	☐ Yes ☐ No	Core plugs properly installed
☐ Yes ☐ No	Retaining straps on core plugs	☐ Yes ☐ No	Crank keys properly installed
☐ Yes ☐ No	Manifold heat-shield holes tapped for 8-32 screws		

PRE-ASSEMBLY FITTING

Measured and Recorded During Pre-Assembly Fitting

☐ Yes ☐ No   Do all valveguides have proper clearance? If no, which are correct _______________________

☐ Yes ☐ No   Do all valveseats meet dimensional specs? If no, which are faulty _______________________

☐ Yes ☐ No   Do all valveseats hold solvent? If no, which leak _______________________

☐ Yes ☐ No   Have all valveguides been machined concentric for press-on seals?

Retainer to Valveguide clearance 0. _______ -inch; adequate on all valves?   ☐ Yes   ☐ No   If no, which valves have insufficient clearance? _______________________

Recommended valvespring seat pressure _______ psi at _______ -inches installed height.

Measured valvespring installed height:

1 _______________   3 _______________   5 _______________   7 _______________

2 _______________   4 _______________   6 _______________   8 _______________

Spring shims used to obtain correct installed height:

1 _______________   3 _______________   5 _______________   7 _______________
2 _______________   4 _______________   6 _______________   8 _______________

Measured valvespring solid height _________-inches

Calculated compressed spring clearance:

1 _______________   3 _______________   5 _______________   7 _______________
2 _______________   4 _______________   6 _______________   8 _______________

Connecting rod bore OK?	☐ Yes	☐ No;	Which rods are defective _______________
Crank straightness OK?	☐ Yes	☐ No	Runout on center main of 0. _________-inch
Main bearing clearance OK?	☐ Yes	☐ No	Measured clearance 0. _________-inch
Crank thrust OK?	☐ Yes	☐ No	Measured clearance 0. _________-inch
Main bearing clearance OK?	☐ Yes	☐ No	Measured clearance 0. _________-inch
Camshaft bearing fit OK?	☐ Yes	☐ No;	Describe problem _______________
Block required clearance grinding for upper sprocket?	☐ Yes	☐ No	
Pin end clearance OK?	☐ Yes	☐ No	Measured clearance 0. _________-inch
Piston-to-wall clearance OK?	☐ Yes	☐ No	Measured clearance 0. _________-inch

Pistons with incorrect clearance _______________

Measured ring end gap:

1 Top ______ 2nd ______   3 Top ______ 2nd ______   5 Top ______ 2nd ______   7 Top ______ 2nd ______
2 Top ______ 2nd ______   4 Top ______ 2nd ______   6 Top ______ 2nd ______   8 Top ______ 2nd ______

Rod bearing clearance OK?	☐ Yes	☐ No	Measured clearance 0. _________-inch
Rod side clearance OK?	☐ Yes	☐ No	Measured clearance 0. _________-inch
Piston-to-head clearance OK?	☐ Yes	☐ No	Measured clearance 0. _________-inch

Cylinders with incorrect clearance _______________

Offset bushings/key used: ☐ +–2°   ☐ +–4°   ☐ +–6°   ☐ +–8°   ☐ +–10°   ☐ +–12°

Rotating assembly clearance OK?	☐ Yes	☐ No;	Cause of interference _______________
Crank index OK?	☐ Yes	☐ No	Maximum ______° out of index on journal no. _______
Cylinder-to-cylinder deck height accurate?	☐ Yes	☐ No	Maximum 0. _______-inch variation.
Rocker geometry OK?	☐ Yes	☐ No;	Describe problem _______________
Rocker-to-stud clearance OK?	☐ Yes	☐ No	Maximum 0. _______-inch (Intake); 0. ______-inch (Exhaust);
Piston-to-valve clearance OK?	☐ Yes	☐ No	Maximum 0. _______-inch (Intake); 0. ______-inch (Exhaust);
Oil pump drive clearance OK?	☐ Yes	☐ No	Measured clearance 0. _________-inch
Intake manifold end-rail clearance OK?	☐ Yes	☐ No	Measured clearance 0. _________-inch
Manifold surface parallel with head?	☐ Yes	☐ No;	Describe problem _______________
Pulleys/accessories aligned?	☐ Yes	☐ No;	Describe problem _______________

____

____

　　FORD FE ENGINES: HOW TO REBUILD